Sustainable Production, Life Cycle Engineering and Management

Series Editors

Prof. Christoph Herrmann
Institut für Werkzeugmaschinen und Fertigungstechnik
Technische Universität Braunschweig
Braunschweig
Germany
E-mail: C.Herrmann@tu-braunschweig.de

Prof. Sami Kara
School of Mechanical & Manufacturing Engineering
The University of New South Wales
Sydney
Australia
E-mail: S.Kara@unsw.edu.au

Joint German-Australian Research Group "Sustainable Manufacturing and Life Cycle Management", www.sustainable-manufacturing.com

For further volumes:
http://www.springer.com/series/10615

Sustainable Production, Life Cycle Engineering and Management

Modern production enables a high standard of living worldwide through products and services. Global responsibility requires a comprehensive integration of sustainable development fostered by new paradigms, innovative technologies, methods and tools as well as business models. Minimizing material and energy usage, adapting material and energy flows to better fit natural process capacities, and changing consumption behaviour are important aspects of future production. A life cycle perspective and an integrated economic, ecological and social evaluation are essential requirements in management and engineering. This series will focus on the issues and latest developments towards sustainability in production based on life cycle thinking.

Sebastian Thiede

Energy Efficiency in Manufacturing Systems

Springer

Author
Dr.-Ing. Dipl.-Wirtsch.-Ing. Sebastian Thiede
Institut für Werkzeugmaschinen und Fertigungstechnik
Technische Universität Braunschweig
Braunschweig
Germany

ISSN 2194-0541 e-ISSN 2194-055X
ISBN 978-3-642-25913-5 e-ISBN 978-3-642-25914-2
DOI 10.1007/978-3-642-25914-2
Springer Heidelberg New York Dordrecht London

Library of Congress Control Number: 2012935578

Printed on acid-free paper

Springer is part of Springer Science+Business Media (www.springer.com)

Foreword

Due to the increased economic and environmental concerns, a systematic consideration of energy and resource consumption is of increasing importance in manufacturing. A realistic and goal-driven analysis and derivation of efficiency potentials demands a holistic system perspective in order to balance conflicting goals and/or to avoid problem shifting. This involves an extended process understanding with all relevant input and output flows and their realistic consumption/emission behavior as well as the necessary consideration of interactions with technical building services. In the field of energy and resource efficiency diverse fields of action need to be distinguished. This could be achieved based on single or continuous data measuring, modeling of energy and resource flows and their interactions as well as appropriate methods for evaluating and predicting machine behaviors. The ultimate objective is to integrate energy and resource oriented variables with the traditional performance indicators (e.g. cost, quality and time) into the decision system of manufacturing companies.

Measures on process and machine level are the first important steps for increasing energy efficiency. However, the consumption of energy and resources and the associated emission of technical equipments are not static but depending on the specific state of operation. On a factory level – which includes coupled interaction of consumers and emitters - individual consumption and emission profiles of processes and process chains lead to certain cumulative profiles for the system as a whole. Thus, in-depth investigation of these consumption and emission profiles on a factory level leads to additional potentials for improving energy efficiency. Due to the dynamic interdependencies within the system, there is a strong demand for a generic energy flow oriented manufacturing simulation environment which would contribute towards improving energy efficiency in manufacturing. The work of Dr. Thiede directly addresses this important topic. With this published work as well as with his active and on-going role, Mr. Thiede has strongly contributed to the development of the Joint German-Australian Research Group "Sustainable Manufacturing and Life Cycle Management" (www.sustainable-manufacturing.com). We are looking forward to continuing our work with Dr. Thiede in future.

Prof. Christoph Herrmann
Technische Universität Braunschweig

Prof. Sami Kara
The University of New South Wales

Acknowledgment

This book was written in context of my work within the Product- and Life-Cycle-Management Research Group of the Institute of Machine Tools and Production Technology (IWF) at Technische Universität Braunschweig. Special thanks go to apl. Prof. Dr.-Ing. Christoph Herrmann as head of the research group for his support of this book as well as the opportunities, freedom and the excellent collaboration I could enjoy while working in the institute.

Furthermore I would like to thank Assoc. Prof. Sami Kara from the Life Cycle Engineering and Management Group of the University of New South Wales (UNSW) in Sydney, Australia, for the fruitful cooperation in context of the Joint German-Australian Research Group "Sustainable Manufacturing and Life Cycle Management" - specifically during my own research stays at the UNSW. My thanks also go to Prof. Dr.-Ing. Prof. h.c. Klaus Dilger and Prof. Dr.-Ing. Thomas Vietor for their contributions which enable the creation of this book.

Big thanks also to all my colleagues in the institute and specifically to those of the Product- and Life-Cycle-Management Research Group. Dear colleagues, thank you very much for the excellent teamwork with many fruitful and nice discussions and experiences which form the positive atmosphere of our team. In particular, I would like to thank Dr.-Ing. Tobias Luger and Dipl.-Wirtsch.-Ing. Tim Heinemann for reviewing the book and their constructive criticism.

Lovely thanks go to my fiancée Jule Schäfer for her understanding and support specifically within the last intensive months when finalizing this book. I thank Janne Schäfer for proofreading. Last but not least, I would like to thank my parents - Annerose and Friedrich-Wilhelm Thiede - for all the freedom and support I got over all the years.

Braunschweig
March 2011

Sebastian Thiede

Contents

List of Figures

List of Tables

List of Symbols and Abbreviations

Symbols

Symbol	typ. Unit	Description
a	hours	scale parameter of Weibull function
b		shape parameter of Weibull function
c_m	kJ/kg K	mass specific heat capacity (e.g. water 4.187)
d	-	effect size / Cohen´s d
ΔT_{FW}	K	temperature difference freshwater - steam
ΔT_C	K	temperature difference condensate - steam
E	kWh	energy (with certain indices)
E_0	kWh	constant energy demand of machine
e_F		constant machine factor
E_S	kW	energy demand for steam generation
F	m^3/s, kg/s	fuel quantity
F_m		manufacturing parameters (e.g. load)
f(t)		failure probability density function
F(t)		failures probability
H	kJ/kg, kJ/m^3	heat/calorific value
h_S	kJ/kg	specific enthalpy steam
h_W	kJ/kg	specific enthalpy water / heat of evaporation
k		machine constant
$\dot{m}_F$	kg/h	fuel consumption
$\dot{m}_S$	kg/h	steam output
MTTF	hours	Mean time to failure
MTTR	hours	Mean time to repair
n_Γ	-	factor of gamma function
n_{FW}	-	share of fresh water for water supply (0..1)
n_C	-	share of condensate for water supply (0..1)
$n_{production}$	pieces	production quantity
n	runs	sample size of simulation experiments
O	-	operation (with indices)
P	W	power
P_{state}	W	power demand for states (e.g. machine - idle, process)
p	bar	compressed air pressure

$\dot{Q}_C$	kW	heat input / combustion capacity
$\dot{Q}_B$	kW	boiler output / boiler capacity
$s_{1..n}$	-	variance of data set 1..n
t	sec	time
t_{state}	sec	duration of states (e.g. idle, process)
T	°C, K	temperature
η_B	%	boiler efficiency
$\dot{V}_F$	m^3/h	fuel consumption
V	m^3	compressed air system volume
$\dot{v}$	m^3/sec	material processing rate
W	J, Ws	work
x	-	values for data set (e.g. output data)

Abbreviations

AE	Auxiliary Energy
ANN	Artificial Neural Networks
BTU	British Thermal Unit
CA	Compressed Air
CBN	Cubic Boron Nitride
CHP	Combined Heat and Power (Cycle)
CNC	Computerised Numerical Control
DCM	Die Casting Machine
DE	Direct Energy
DES	Discrete Event Simulation
EMIS	Energy Management Information System
EnMS	Energy Management System
EPE	Embodied Product Energy
ERP	Enterprise Resource Planning
EU	European Union
EV	Evaluation and Visualisation (module)
FEM	Finite Element Method
GHG	Green House Gas
CIRP	College International pour la Recherche en Productique/ The International Academy for Production Engineering
ICT	Information and Communication Technology
IE	Indirect Energy
ISO	International Organisation for Standardisation
IWF	Institute of Machine Tools and Production Technology, TU Braunschweig
LCA	Life Cycle Assessment
LCC	Life Cycle Costing
LCI	Life Cycle Inventory
MCDM	Multi Criteria Decision Making
MLE	Maximum Likelihood Estimation
MRR	Median Rank Regression
MS	Manufacturing/Main System (module)
MTTR	Mean Time to Repair
MTTF	Mean Time to Failure
OR	Operations Research
PCB	Printed Circuit Boards
PDCA	Plan Do Check Act
PLM	Product Lifecycle Management
PM	Process Module
PPC	Production Planning and Control
VSM	Value Stream Mapping
SMD	Surface-Mounted Device
SME	Small and Medium sized enterprises

STD	Standard Deviation
TBS	Technical Building Services
TE	Theoretical Energy
TEEM	Total Energy Efficiency Management
THT	Through Hole Technology

Chapter 1
Introduction

1.1 Sustainability as New Paradigm in Manufacturing

Nowadays manufacturing companies are facing diverse economic (e.g. shorter product life cycles, rising product variant diversity, increasing production volume fluctuations, rapid changing technologies, financial crisis) but also enormous environmental (e.g. climate change, resource depletion) and social challenges (e.g. aging personnel).

Especially the attention to environmental aspects like global warming or resource depletion is accelerating and different drivers are exerting pressure on companies (Figure 1). It is more and more an issue **addressed in politics** (e.g. EU 2020 climate goals) and rising public awareness - potentially resulting in challenging **consequences on the corporate image** - can be observed. In addition, drivers like **increasing energy and raw material prices**, the potential **lack of critical resources**, necessary **investments** for environmental sound technologies, and **penalties for lacking compliance** with environmental regulations as well as regulative **incentives** or the introduction of CO_2 certificates are issues that directly connect environmental driven issues to business objectives of a company.

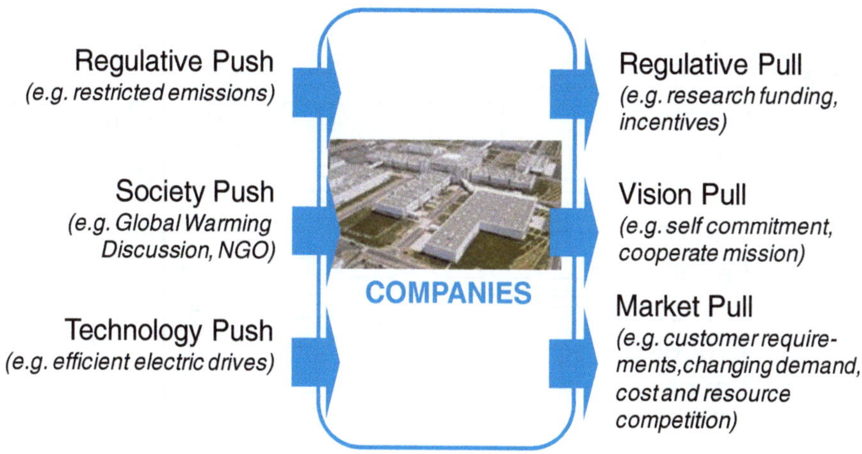

Fig. 1 Drivers for sustainability in manufacturing companies (adapted from Fichter, 2005)

S. Thiede: Energy Efficiency in Manufacturing Systems, SPLCEM, pp. 1–8.
springerlink.com

Therefore, besides traditional economical production objectives (e.g. cost, time, quality), environmental driven objectives (e.g. low CO_2 emissions) have become strategically relevant for manufacturing companies. Altogether, it is necessary to strive for harmonising the requirements of a sustainable development with the needs of manufacturing (Brundtland Commission, 1987). Manufacturing processes play an essential role regarding economic success and environmental impact. Production processes consume raw materials and transform them into products and wanted or unwanted by-products using energy as input. While one part of the resources is used for creating value and embodied into the form and composition of products, another part is wasted in terms of losses, heat and emissions. Manufacturing systems predominantly influence the environmental outcome and therefore represent the major potential to minimise the environmental performance of a company (Warnecke et al., 1998). Thus, designing and improving manufacturing systems while advantageously integrating economic, ecological and social goals becomes an essential strategic objective of manufacturing companies nowadays (Herrmann, 2009; European Commission, 2006; Schultz, 2002). It is clear, that an isolated consideration of traditional economic variables is not sufficient anymore. In fact, **Sustainable Manufacturing is the new necessary paradigm for manufacturing companies** which involves the integration of all relevant dimensions for all technological and organisational measures within the normative, strategic and operative production management (Figure 2).

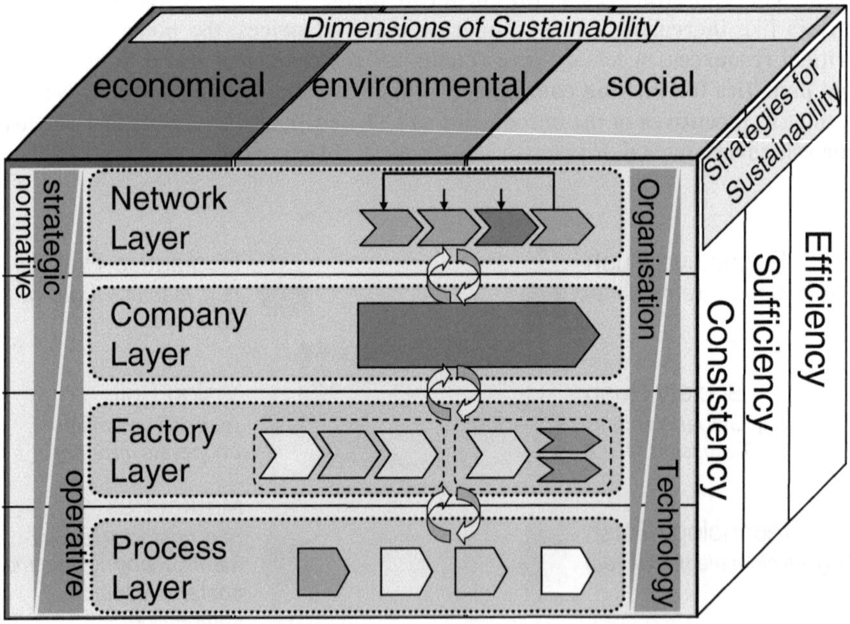

Fig. 2 Framework for Sustainable Manufacturing (Herrmann, 2009; Herrmann et al., 2008a)

Therefore, all technological and organisational measures within manufacturing companies have to be evaluated based on a comprehensive set of criteria nowadays which involves the integration of the economic, environmental and also the social perspective (known as the triple bottom line). As a holistic approach which strives to avoid problem shifting within manufacturing companies, their supply chain and life cycle phases, this involves the consideration of all basic strategies of sustainability on different layers beginning from the single (production) process, process chains on a factory layer, strategic decisions on a company layer or activities in closed looped supply chains like utilising Re-X-options, such as remanufacturing or refurbishment (network layer) (Herrmann, 2009). In Figure 3 the strategies of a sustainable development are depicted based on the coherence of economic and environmental impact. While efficiency strives to minimise the material and energy usage in all life cycle phases by increasing resource productivity, sufficiency demands a change in the behaviour of usage and consumption. The third strategy of sustainability is consistency, which can be defined as the adaptation of material and energy flows to fit adequately to biological process capacities (Dyckhoff and Souren, 2008; Herrmann et al., 2007a; Herrmann, 2009).

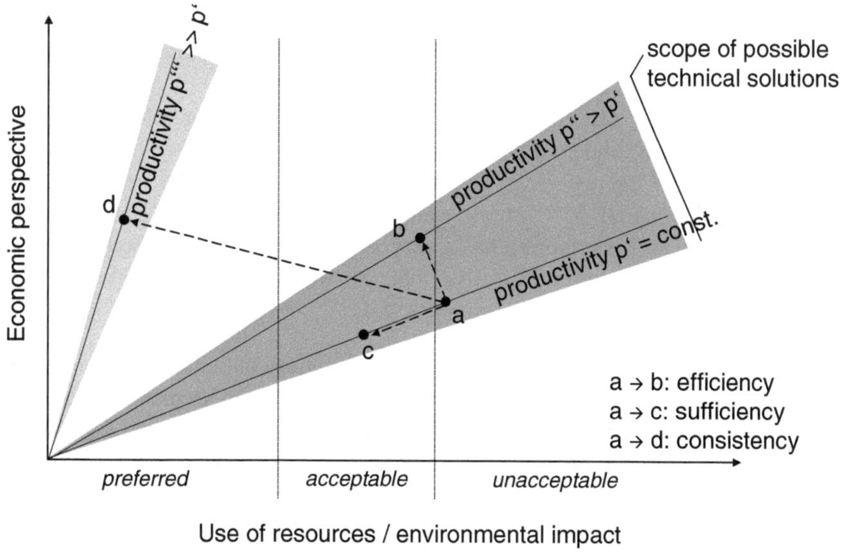

Fig. 3 Strategies for a sustainable development (Schmidt, 2007)

As shown in Figure 3 the *sufficiency* strategy may involve the conscious reduction of (economic) growth, which impedes a broad application in companies. Significant improvements in sustainability can be achieved by the preferential application of the strategy *consistency*, since this forces the substitution of processes, with which the potential environmental impacts are minimised and harmful materials are avoided. While having significant improvement potential this strategy involves certain development and implementation efforts in terms of time and costs. Up to now typically from an economic (cost) perspective, *efficiency* as improvement of the output to input ratio is an established strategy in companies already. It also bears significant potential in terms of environmental improvement and enables a decoupling of economic growth and related environmental impact. However, the application of the strategy efficiency can result in rebound effects, which have to be taken into account. Therefore, in order to consider all interdependencies in advance to the implementation of strategies, a holistic perspective on the considered system as well as an appropriate methodology is of importance.

1.2 Motivation

Within the broad paradigm of sustainable manufacturing, the issue of energy efficiency will be addressed specifically in this book. It focuses on increasing the efficiency of energy flows in manufacturing companies with certain impact on both economic as well as environmental target variables. This automatically includes an improvement of resource efficiency as well since these energy flows are typically directly or indirectly connected with the depletion of critical resources (e.g. oil, gas, coal).

The topic "energy efficiency in manufacturing" is of major relevance from a national as well as a single company perspective. On a national scale, industry is a major consumer of energy – e.g. German industry is responsible for 42% of the national electricity and 35% of the national gas consumption (BMWi, 2011). Considering energy consumption has a very strong relevance from both economic as well as environmental perspective. On the one hand the energy supply is directly connected with ecological impacts, e.g.:

- Green house gas (GHG) emissions with significant contribution to global warming. As an example, only through energy demand industry is responsible for approx. 28% of CO_2 emissions (plus approx. 9% through direct industrial emissions, see Figure 4) in Germany (BMWi, 2011).
- Depletion of diverse non-renewable resources (e.g. oil, gas, coal) with possible lack of these resources in the future - based on currently known securely mineable deposits and demand the statically estimated supply range is approx. 40 (oil) respectively 60 (gas) years (BMWi, 2011).
- Risks and consequences of using nuclear power plants for electricity generation such as possible hazardous accidents with nuclear pollution and problem of radioactive waste disposal.
- Land use and harm to landscape and biodiversity through e.g. mining of coal, oil or uranium or installation of e.g. wind energy equipment.

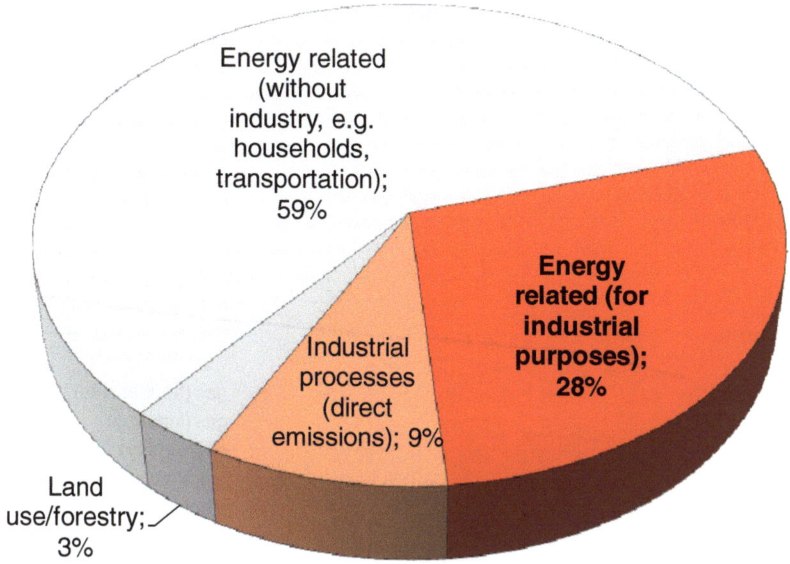

Fig. 4 Composition of CO_2 emissions for Germany (BMWi, 2011)

On the other hand, energy consumption also has a very strong economic dimension. Energy prices for electricity, gas and oil are disproportionately and steadily increasing in the last years (Figure 5). As a result, energy costs can make up a very relevant share on total costs of manufacturing companies today. Studies estimate that energy costs may sum up to 20% on total costs (in some branches) – the average for manufacturing companies is approx. 6% nowadays (Thamling et al., 2010; IHK 2009). An increase to an average share of approx. 8% is expected until 2013 (IHK 2009).

Recent studies driven from research as well as industrial practice also underline the importance of energy efficiency in manufacturing. In an industry survey with SME (small and medium sized enterprises) approx. 70% named energy efficiency as an important topic. The main motivation is clearly to decrease energy costs whereas also the contribution towards environmental protection is an important reason (Thamling et al., 2010). However, studies also underline the unemployed potential regarding energy efficiency in manufacturing as well as obstacles which impede an identification and broad applicability of improvement measures in practice (Schröter et al., 2009). Obviously there is a strong need of appropriate methods and tools to support fostering energy efficiency in manufacturing companies.

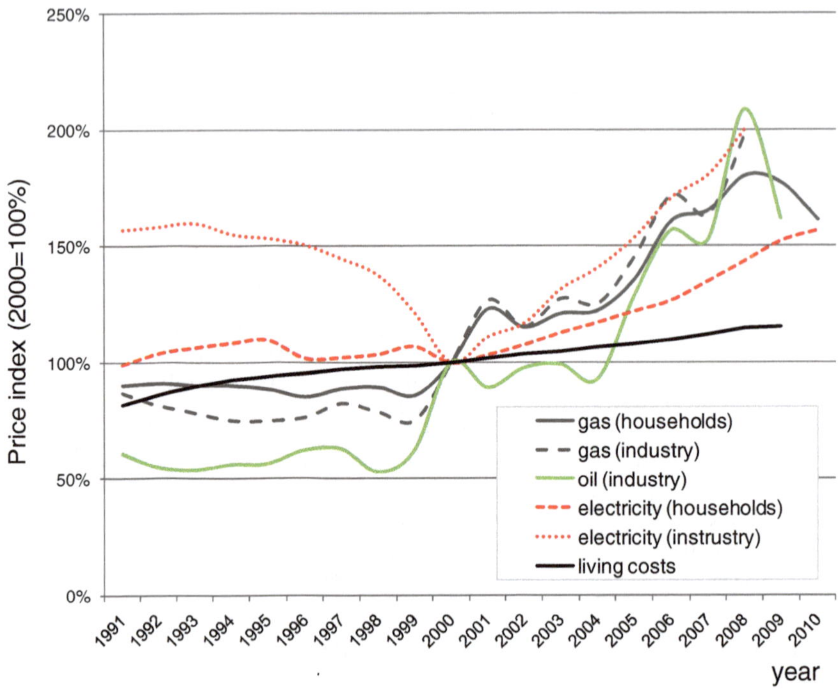

Fig. 5 Development of energy prices in Germany (compared to progression of standard living costs) (BMWi, 2011)

1.3 Objectives and Work Structure

Against the described background as main objective this book aims at

> Strongly **contributing towards the improvement of energy efficiency** in manufacturing.

The structure of the book is shown in Figure 6. Following this introduction the necessary technical background in context of manufacturing and related energy consumption will be given (Chapter 2). Based on this as well as industrial experiences, diverse requirements will be derived which serve as background for reasoning the methodological approach taken here (Chapter 3). These methodological considerations formulate the objective of

> Developing an **energy flow oriented manufacturing system simulation** approach.

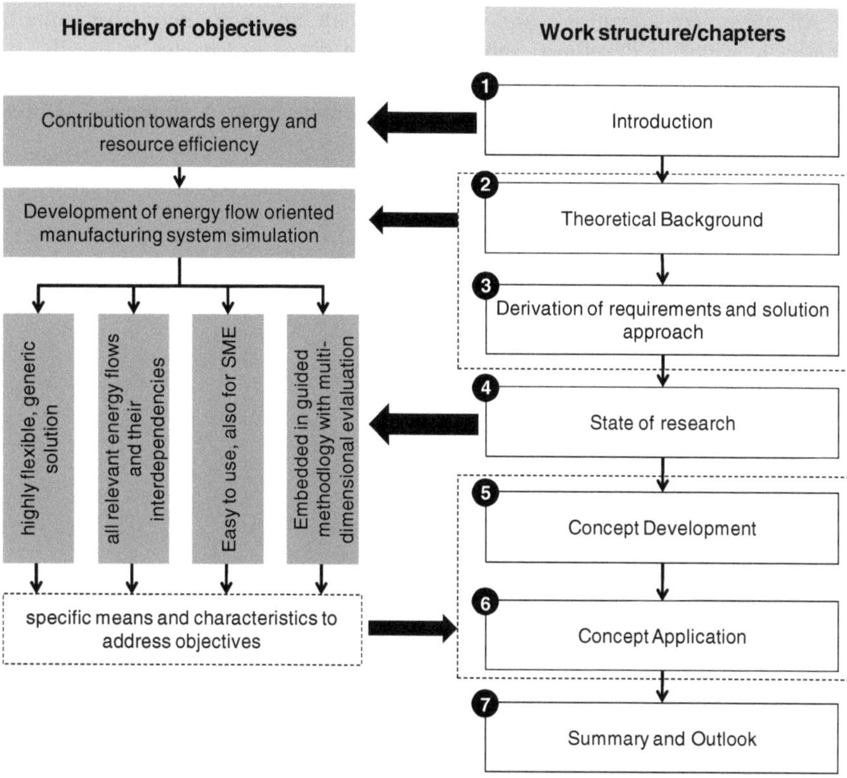

Fig. 6 Hierarchy of objectives and related structure of the book

In the next step, the more general requirements are broken down to very specific criteria afterwards. With that, relevant available research approaches are being analysed and evaluated in detail in order to derive necessary further research demand (Chapter 4). Based on this detailed analysis, further specific objectives can be identified.

The aim is to develop an energy flow oriented manufacturing system simulation approach which

- is **not related or restricted to a specific case but generic in nature and applicable to manifold production situations** in the sense of a generic simulation environment.
- explicitly pursues a holistic perspective including **all relevant energy flows as well as their interdependencies**.
- is **also applicable for small and medium sized enterprises** typically facing obstacles towards energy efficiency measures and usage of simulation.
- is **embedded in a guided methodology** for goal-oriented identification and realistic as well as **multi-dimensional evaluation** of improvement measures in all relevant fields of actions.

All these considerations are incorporated in an own innovative solution approach, which is developed and explained in detail in Chapter 5. Finally, the flexible applicability and potentials of the approach are shown in four different case studies (Chapter 6) before closing the book with a summary, concept evaluation and an outlook (Chapter 7).

Chapter 2
Theoretical Background

Against the background of the scope and objectives of the planned research work, the following chapter will provide the necessary theoretical background. First of all the basics of manufacturing and energy consumption will be presented. Following this, the state of art regarding energy efficiency measures in manufacturing is described which serves as base for deriving requirements and potentials for further research demand.

2.1 Production and Production Management

In the field of production engineering and management a wide range of different terms and synonyms are – not always consistently - used in different disciplines in research and industrial practice. In order to ensure a necessary and mutual understanding basic definitions and the connected theoretical background will be given as base for this book. As far as possible, the terminology will reflect the glossary/dictionary of the CIRP, The International Academy for Production Engineering (C. I. R. P., 2008; C. I. R. P., 2004a, C. I. R. P., 2004b).

As a very general term, **Operations Management** "deals with the design and management of products, processes, services and supply chains. It considers the acquisition, development, and utilisation of resources" which companies transform into "the goods and services their clients want" (Massachusetts Institute of Technology (MIT), 2010). Whereas this definition is relatively broad and includes all types of transformation and value creation in a company, production as a part of it is focusing on physical transformation into tangible results. **Production** can be defined as a combination of production factors such as labour, material and technical equipment for the purpose of value creation in form of products (Gutenberg, 1983). Still the term production is relatively broad in nature and can also be applied for other areas like the agricultural sector or service industry (intangible products), which are not the main focus of this book. Thus, the term **Manufacturing** is also used which is more specifically "the business or industry of producing goods in large quantities in factories [...]" (Oxford University Press, 2011). In literature, there is a certain inconsistency regarding the usage of those terms; in this book both expressions are used while "production" is larger and includes "manufacturing" but not vice versa.

S. Thiede: Energy Efficiency in Manufacturing Systems, SPLCEM, pp. 9–34.
springerlink.com

Figure 7 underlines the understanding of production in context of manufacturing as transformation of inputs like

- raw material (e.g. steel),
- auxiliary and operating material (e.g. coolants, paint, screws),
- energy (e.g. electricity),
- labour/personnel (e.g. for operating and maintaining the machine),
- technical equipment for main production process and supporting processes (e.g. transport, storage, measuring),
- information,

into wanted (valuable products) and unwanted (scrap, waste, exhaust heat/air) outputs (Westkämper, 2005; Schenk, 2004). It also shows a possible classification of production related transformation processes based on German standard DIN 8580. The actual embodiment of production processes is typically called **Production Engineering**.

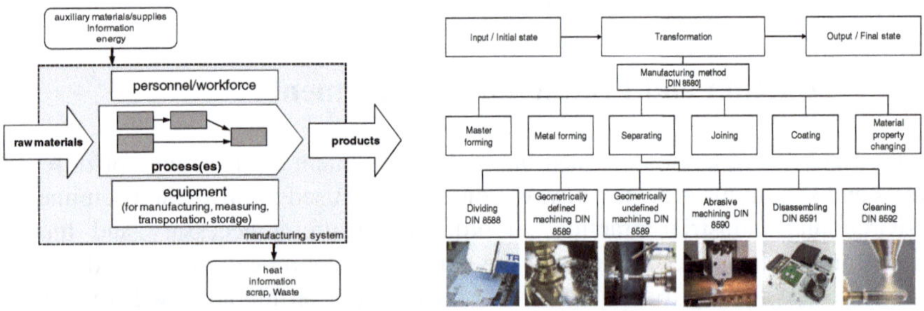

Fig. 7 Production as Transformation from Inputs into Outputs (Westkämper, 2005; DIN 8580)

Like any other process, a production process is a "set of interrelated activities [value creating and supporting activities like transformation, combination, transport, control, measure or storage (Barbian, 2005)] which transforms inputs into outputs" whereas the "inputs to a process are generally outputs of other processes" (DIN 9000). Complex technical products are typically made in multi-step production process chains as logically linked sequence of successive or parallel single processes (and associated activities) over time with one common goal namely to bring out a defined output (one or several final products) at the very end (e.g. Arnold, 2002). These processes and process chains involve technical equipment and personnel, which form manufacturing systems as specific designated areas for production and, on a higher level of aggregation, factories (Figure 8).

In this context manufacturing systems can be classified according to different criteria, which specify the properties of the specific system (Figure 9).

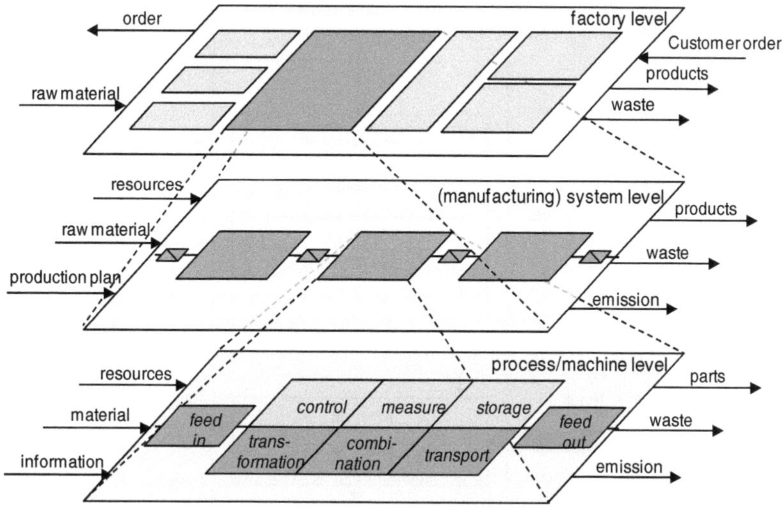

Fig. 8 Levels of abstractions in production/manufacturing (Herrmann et al., 2007b based on Barbian, 2005)

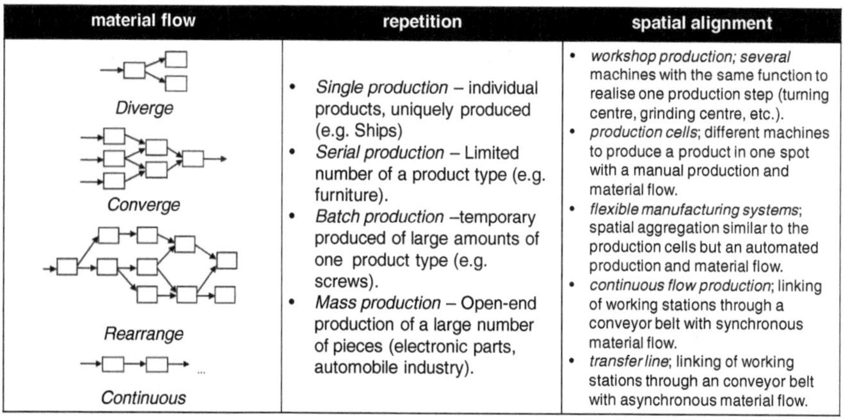

Fig. 9 Classification of manufacturing systems (e.g. Dyckhoff and Spengler, 2010; Schuh, 2006; Westkämper, 2005)

By common definition **Production Management** is responsible for planning and controlling production in order to produce “the right product in terms of type and quantity, in the right quality, at the right time and, for acceptable costs.” (e.g. Westkämper, 2005) Figure 10 shows the connected control loop of production management. As also mentioned in the definition, main reference input variables of production management typically refer to costs, time (e.g. reliability, speed) and quality targets (e.g. Bickford et al., 1996). Production management can

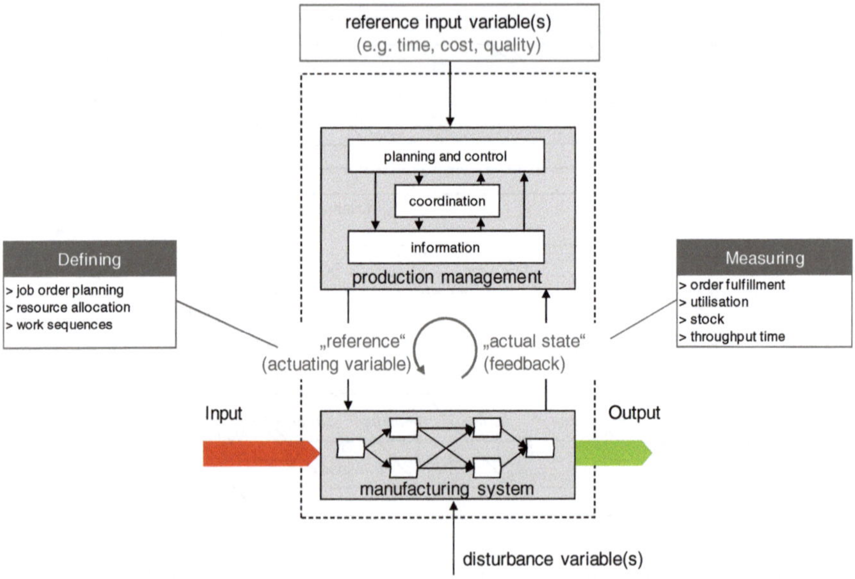

Fig. 10 Control loop of production management (Dyckhoff and Spengler, 2010; Dyckhoff, 1994)

influence the manufacturing system through actuating variables on a strategic (e.g. production structure/layout) and operative (e.g. job order planning, resource allocation) layer. Feedback variables (e.g. utilisation, throughput times) enable the comparison of the reference with the actual state, which might differ due to disturbance variables acting on the manufacturing system. The control loop is closed through adjusting actuating variables in order to meet the objectives management (Dyckhoff and Spengler, 2010).

2.2 Energy and Energy Supply

By popular definition "energy is the capacity to do work" (e.g. McKinney et al., 2007) respectively "the inherent ability of a system to generate external impact" (e.g. Planck and Päsler, 1964) – therefore it is necessary to execute any kind of designated tasks. Energy (E) is a state variable connected with Work (W) as process variable, which describes the energetic difference when a system changes from one state to the other. Power (P) is the rate of energy usage related to a period of time (t).

$$W = P * t \tag{1}$$

$$P = \frac{\Delta W}{\Delta t} \tag{2}$$

The standard unit (derived from SI-units) for energy is Joule [J], for power it is Watt [W].

$$1\,J = 1\,Nm = 1\,W * s = 1\,\frac{kg\,m^2}{s^2} \qquad (3)$$

However, for different areas of application diverse units for energy can be found. The conversion between different energy units is shown in Figure 11.

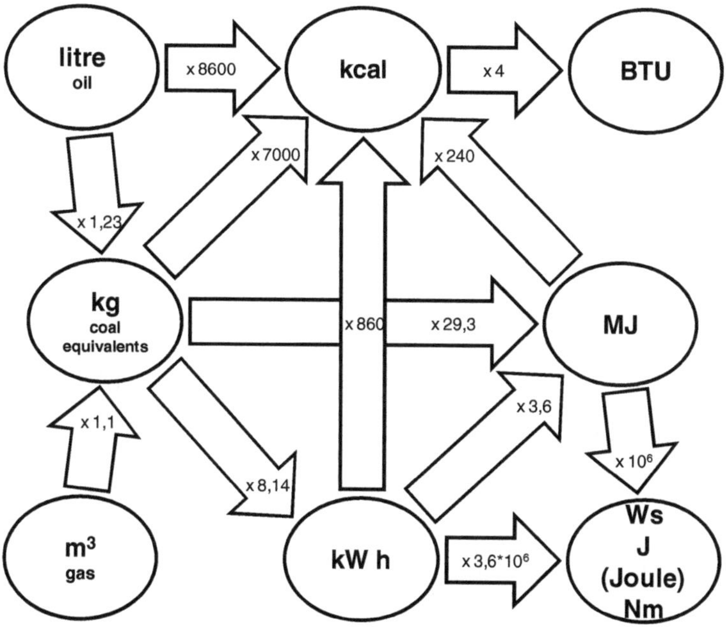

Fig. 11 Conversion between popular energy units (Dehli, 1998)

Fundamental physics distinguish between only two basic types of energy: potential (stored) and kinetic (working) energy (Viegas, 2005; EIA, 2009). However, when going into more details with mechanical, thermal, chemical, electric, electromagnetic and nuclear energy more forms can be differentiated (e.g. EIA, 2009). Conversion between different energy forms is basically possible. Referring to the two basic laws of thermodynamics within a system the sum of energy stays constant but every conversion is connected with losses because not the whole amount of energy ends in the designated form. In this context energy is considered as sum of exergy and anergy: exergy is the usable part of energy of a system, which is being converted from one energy form to the other. Anergy is energy which cannot be further utilised and is referred to as loss (typically in form of heat) (Müller et al., 2009). A system strives towards a share of exergy of zero, which means that it is in equilibrium and no further work can be done.

To enable (technical) application of energy, conversion in between different forms is inevitable. Figure 12 shows selected conversions with connected efficiencies resulting in certain losses.

transformation from	in	through	efficiency
electrical	electrical	transformator	95%
	mechanical	electric drive	95%
	thermal	electric heating	100%
	chemical	battery, electrolysis	70%
		light bulb	5%
	radiation	fluorescent lamp	20%
		laser	up to 35%
mechanical	electrical	generator	95%
	mechanical	gearbox	99%
	thermal	mech. brake	100%
thermal	electrical	thermocouple	5%
	mechanical	diesel engine	35%
		otto engine	25%
	thermal	heat exchanger	90%
chemical	electrical	Battery	5%
		fuel cell	35%
	thermal	coal heating	25%

Fig. 12 Efficiency of selected energy conversion processes (Müller et al., 2009)

In an energy supply chain different energy carriers are of importance (Figure 13). In nature, primary energy - without any conversion so far - can be found in form of e.g. oil, gas, coal (chemical energy) or renewable source (e.g. radiation/solar energy). These primary energy carriers are being converted to secondary energy (e.g. electricity, heating/fuel oil) and transferred to the designated destination. Further conversions into the targeted/useful form of energy (e.g. compressed air, heat/cold) might be necessary in order to fulfil the designated function (e.g. enable rotation of drives, movement of actuators, heating up space) (Brettar, 1988; VDI, 2003). Against the background of the physical coherences as described before, this whole supply chain involves losses from conversions itself and transmission as well as inappropriate control and usage (e.g. leakages). For example, in Europe (on average) electricity has a primary energy factor of about 3.3 - that means for each kWh of electricity 3.3 kWh of primary energy need to be deployed (ISO EN15603).

Figure 14 shows the energy flows from a nation's perspective, in this case Scotland. It reveals a typical mix of energy sources for electricity generation and the significant amount of energy, which is involved as well as the main consumers of different forms of energy.

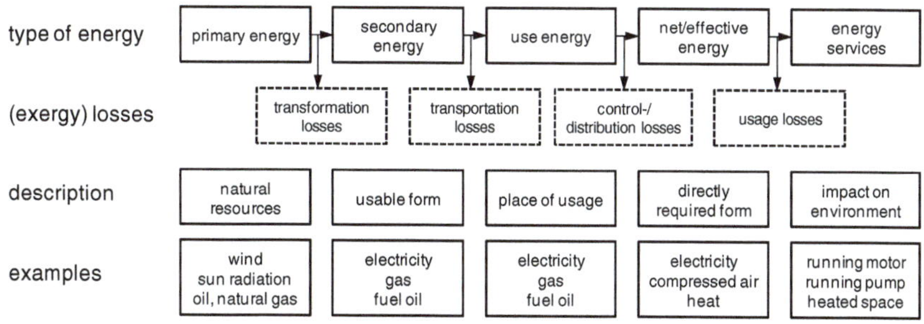

Fig. 13 Energy supply chain (Engelmann, 2009)

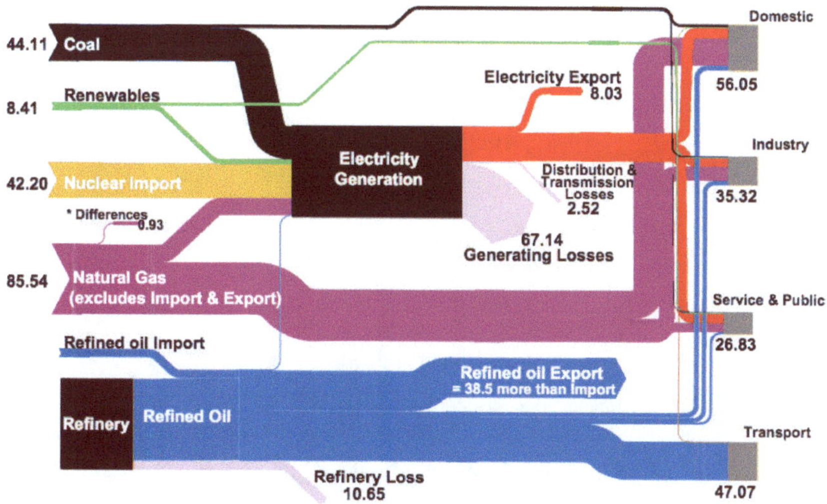

Fig. 14 Energy flow diagram for Scotland (Scottish government, 2006)

The supply with energy is directly connected with environmental impacts. On the one hand energy consumption involves the depletion of diverse non-renewable resources (e.g. oil, gas, coal). Besides issues related to the actual exploration of these resources (e.g. mining), this is a challenge in the longer-term perspective: based on currently known securely mineable deposits and demand the statically estimated supply range is approx. 40 (oil) respectively 60 (gas) years (BMWi, 2011). On the other hand, the generation and usage of energy through burning coal, gas or oil results in green house gas (GHG) emissions with significant contribution to global warming. GHG emissions from electricity usage directly depend on the actual mix of energy sources for generation, which strongly differs between countries. Generally, three different energy sources can be distinguished:

- Conventional thermal energy generation by incineration of non-renewable resources such as coal or gas.

- Nuclear power generation.
- Energy generation from renewable resources, such as wind, water or solar power.

Figure 15 shows the energy mix composition for the electricity net generation in different countries worldwide. Significant differences can be observed between countries largely depending on conventional thermal energy generation with high specific GHG emissions, such as Australia (0.924 kg CO_2/kWh electricity, EIA, 2009) or Saudi Arabia (0.816 kg CO_2/kWh electricity), and countries mainly relying on renewable energy sources like Brazil (0.093 kg CO_2/kWh electricity) or Norway (0.005 kg CO_2/kWh electricity). Thus, energy consumption in specific countries is associated with a specific environmental impact depending on the sources.

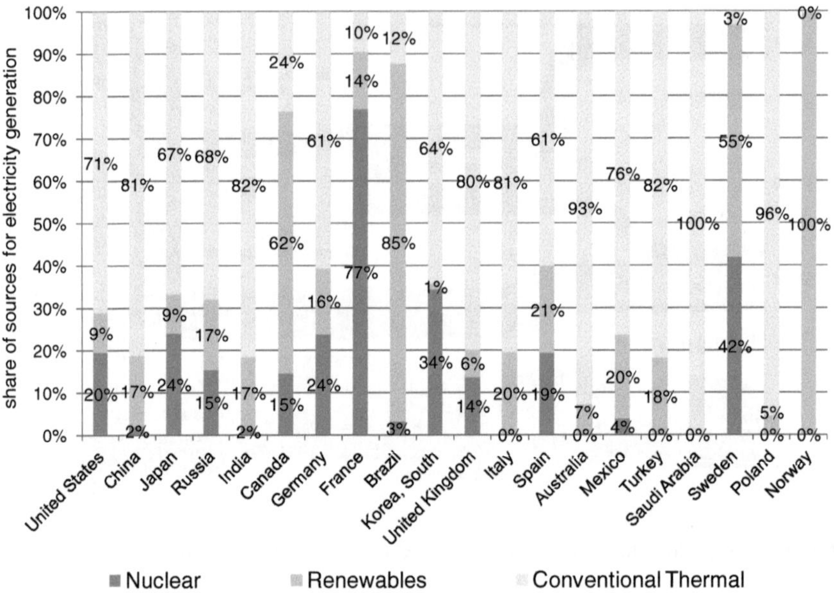

Fig. 15 Electricity net generation 2008 by type and country (top 20 countries) (EIA, 2009)

2.3 Energy Consumption in Manufacturing

2.3.1 Forms of Energy Consumption in Manufacturing

As described before, manufacturing processes require a significant amount of resources and energy whereas one part of the input is used for creating value, another part is wasted in terms of losses. Hence, it is involving relevant (and to a certain extend unavoidable) environmental impact through energy consumption with related resource depletion and GHG emissions. Table 1 shows the necessary forms of energy for industrial (manufacturing) purposes in the case of Germany.

According to that, (space and process) heat and mechanical energy are mainly needed (Seefeldt and Wünsch, 2007) which are getting converted from energy sources like electricity (electrical energy), gas, oil or coal (chemical energy). The study also underlines that the actual composition of energy form and sources differs significantly between different branches. Whereas coal is mainly used in metal founding, cement or chemical industry (almost 90% of coal is used by these branches), oil and especially electricity as well as gas are far more common through all other industries. In machinery and automotive industry for example, electricity counts up for over 50% of total energy consumption (Seefeldt and Wünsch, 2007).

Table 1 Energy consumption for German producing industry with respect to energy forms and sources (based on data from 2002, in Petajoule)

	space heat	process heat	mechanical energy	lighting	total
total	**345.6**	**1589.3**	**522.3**	**72.1**	**2529.3**
electricity	21.8	234.8	490.4	72.1	**819.1**
gas	179.6	792.1	2.0	0.0	**973.7**
oil	97.7	129.6	3.9	0.0	**231.2**
coal	10.3	397.5	0.0	0.0	**407.7**
district heat	27.8	27.9	0.0	0.0	**55.7**
renewable	8.4	7.4	0.0	0.0	**15.7**
fuel	0.0	0.0	26.1	0.0	**26.1**

On a national scale, industry is one of the major consumers of natural gas as primary energy carrier, e.g. in Germany the share is 36% (BMWi, 2011). Additionally, industry consumes the major share of electricity which is a secondary energy carrier and is produced using primary sources including significant losses. In Germany, industry is responsible for the consumption of 47% of the national electricity (BMWi, 2011). As mentioned above, energy consumption has a very strong relevance from both an economic as well as an environmental perspective. Thereby the pure energetic view as shown in Table 1 is only one perspective; whereas striving towards sustainability in manufacturing demands a more detailed analysis of connected economic as well as environmental impacts (here depicted with related CO_2 emissions). Therefore (based on the data from Seefeldt and Wünsch, 2007) Table 2 and related Figure 16 show the estimated energy costs and CO_2 emissions of the German manufacturing industry for the main energy sources.

The calculation is based on the average energy prices for the considered years and the emitted CO_2 for either generating electricity (energy source mix for Germany) or directly burning oil, gas or coal. The calculations underline the major importance of

Table 2 Energy consumption of manufacturing companies with related costs and CO_2 emissions (for Germany)

	energy consumption	energy costs (2000)	energy costs (2008)	related CO_2 emissions
	[in PJ]	[in €]	[in €]	[in t]
electricity	819,1	10.012.650.793 €	20.073.221.336 €	130.933.135
gas	973,7	4.577.253.331 €	9.094.440.438 €	38.745.623
oil	231,2	1.055.855.319 €	2.204.659.000 €	10.395.556
coal	407,8	586.200.977 €	1.566.545.164 €	37.185.949
total	**2431,8**	**16.231.960.420 €**	**32.938.865.938 €**	**217.260.264**

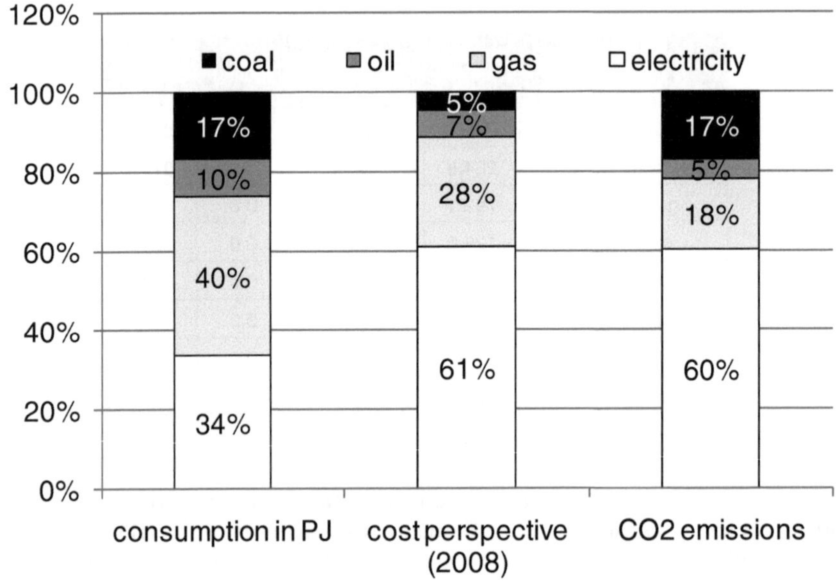

Fig. 16 Estimation of costs and CO_2 emission related to energy consumption of German manufacturing companies

considering electricity in comparison to primary energy sources (due to upstream supply chain). Only through its electricity consumption, industry is responsible for approx. 18% of CO_2 emissions (plus approx. 20% through direct industrial emissions) in Germany (BMWi, 2011). Furthermore, the calculation stresses the very strong economic relevance of industrial energy consumption. Energy prices for electricity, gas and oil have been steadily increasing for the last couple of years (BMWi, 2011). As shown in Table 2, energy costs for manufacturing companies have been more than doubled from the year 2000 to 2008.

2.3.2 Consumers of Energy

Table 2 already gave an overview over main energy forms needed in industry. Altogether the most typical energy conversions are from gas to process heat and from electricity to mechanical energy. Due to their relevance in general and for this book in particular, selected energy flows from these categories will be presented in more detail in chapter 2.4. For a deeper insight regarding the coherences in a manufacturing company, Figure 17 shows internal energy flows with respect to different consumers and energy carriers. The figure underlines the manifold technologies, which are involved to keep a factory operating whereas the actual embodiment evidently depends on the specific case. On average, space and process heat sum up to a major share on total energy consumption (in PJ or kWh) in industry, altogether approx. 75%. However, this consumption mostly bases on gas, coal or oil and is also branch specific. As shown above, electricity is of specific relevance due to its cost as well as environmental impact and the broad range of application. Therefore, Figure 17 shows typical users of electricity in industry. It is mainly used to run electric drives to generate mechanical energy. Typical applications are pumps, air conditioning (chill generation, ventilation), compressed air generation and of course the actual movement and processing of production machines (e.g. spindle motor, conveyor belt drive). Furthermore electricity is necessary to operate lighting as well as information and communication technologies (ICT) (Schmid and Layer, 2003).

This consideration focuses on cross-sectional technologies with broad relevance for all industries to give a general overview from an energetic perspective. In the

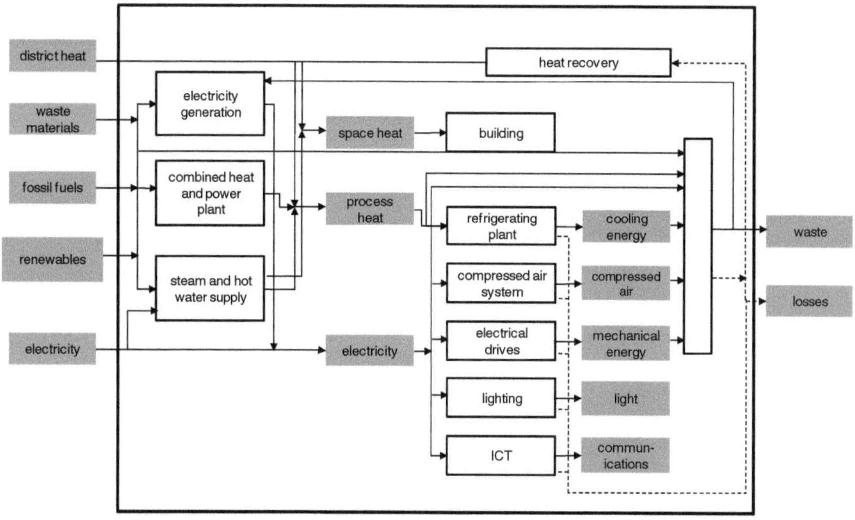

Fig. 17 Internal energy consumers and flows in a manufacturing company (Schmid, 2008)

specific case, these technologies are applied in very complex production environments and embodied in specific machines. Energy consumption takes place on diverse levels of consideration: it can be distinguished between processes and machines for actual value creation and energy consuming equipment for diverse supporting activities (e.g. coolant treatment in machining Bode, 2007) including building shell and technical infrastructure (Schenk, 2004; Clarke et al., 2008). In this context, the term (technical) building services (TBS) is often used. TBS are responsible for essential tasks like heating and cooling (e.g. space and process heat), ventilation and air conditioning (e.g. exhaust air purification, air technology), power engineering (e.g. energy supply, lighting), or water/media supply and treatment. Hence they provide the needed production environment and necessary process energy in different forms as well as process-related media like water. (Hall and Greeno, 2009; Chadderton, 2004). Altogether, referring to a European study, 35-40% of industry's energy consumption is caused by TBS (Eichhammer et al., 1996). Altogether, an example breakdown of different energy consumers in a factory is shown in Figure 18.

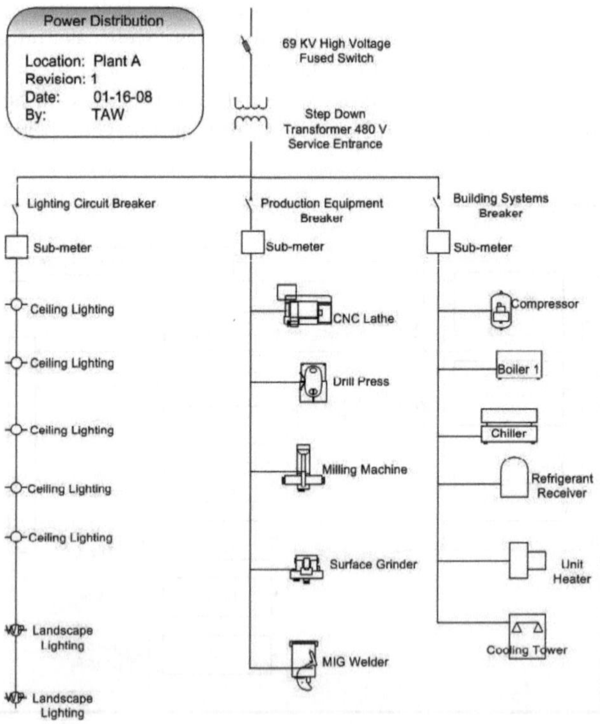

Fig. 18 Simplified structure of energy (here: electricity) consumers in a factory (Westerkamp, 2008)

2.3.3 Energy Consumption Behaviour of Production Machines

In addition to the general overview of energy consumption in manufacturing companies, the analysis of the consumption behaviour of production machines is necessary. As diverse studies for different types of production machines show, their energy consumption is usually not constant over time but rather highly dynamic depending on the production process and the actual state of the machine. Machines consist of several energy consuming components (e.g. electric drives) that generate a specific energy load profile when producing (Eckebrecht, 2000; Gutowski et al., 2006; Binding, 1988). This typically applies to electricity, but is also true for other forms of energy like compressed air, process heat or gas since their consumption naturally also differs depending on process and machine states. As example, Figure 19 shows an electrical load profile for the case of a grinding machine.

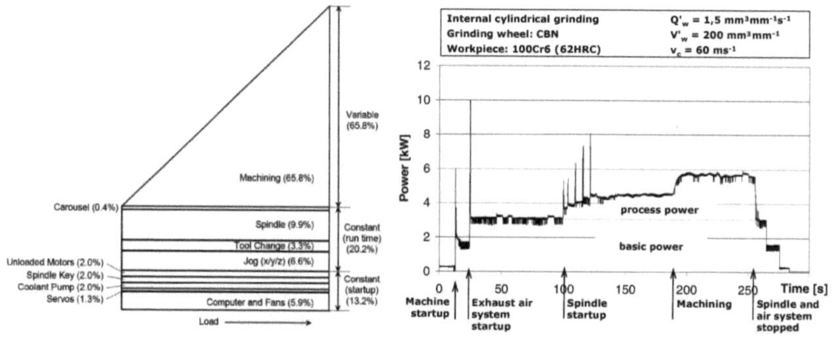

Fig. 19 Energy used as a function of material removal rate for a 3-axis CNC milling machine (left, from Gutowski et al., 2006) and electrical energy consumption of a grinding process (excluding filter system) (Herrmann et al., 2008b)

In general, different typical main states of a machine can be distinguished, whereas, depending on the specific machine, a more detailed differentiation or combination of states is possible (e.g. Binding, 1988; Dietmair and Verl, 2008; Dahmus and Gutowski, 2004; Devoldere et al., 2007):

- **Off:** main switch off, no energy consumption
- **Start-up:** many machines consist of distinctive start-up phases, with energy demand peaks caused by switching on certain components, heating-up phases etc.
- **Idle:** typically relatively constant energy consumption after main supporting components completed start-up and machine is "ready for production".
- **Run-time/ready for machining:** positioning and loading straight before actual processing (e.g. movement of spindle in position towards workpiece but without material removal)
- **Operation:** actual production process takes place, physically necessary energy to fulfil production task (e.g. remove material)

In general, energy profiles can be subdivided into constant and variable energy consumption (Figure 19, Gutowski et al., 2006). The constant energy consumption includes the energy requirements of machine components like control units, pumps (e.g. oil pressure, coolant) or coolers, which enable an operating state. The variable energy consumption of a production machine enfolds the required energy for tool handling, positioning and the actual operation (e.g. cutting). Studies have shown that machine tools with increasing levels of automation reveal higher constant energy consumptions resulting from the amount of additional integrated machine components. The energy consumption E is therefore not only determined by the cutting operation, but may be dominated by the basic power consuming components (Klocke et al., 2010; Dahmus and Gutowski, 2004). Altogether, the energy consumption of the machine as a whole (E) depends on the design and control schemes of the machine as well as actual process parameters. A simplified linear consumption equation can generally be formulated as (Wohinz and Moor, 1989):

$$E = E_0 + e_F{*}F_m \tag{4}$$

In this case E_0 is the constant energy demand (typically idle demand) whereas the term $e_F{*}F_m$ describes the variable demand depending on constant factors reflecting machine characteristics (e_F) and the actual manufacturing parameters (F_m, e.g. load, speed, production volume). Similar to this equation, Gutowski et al. specify for the case of electricity demand of a machine tool (Gutowski et al., 2006):

$$P = P_{idle} + k{*}\dot{v} \tag{5}$$

P [in kW] is the total power consisting of P_{idle} (idle power), a machine specific constant k and the rate of material processing $\dot{v}$ in cm^3/sec.

A certain period of production time typically involves both times of operation and machine idleness. Having in mind equation (1) and (4)/(5) the following equation for the energy consumption (of e.g. one shift) can be formulated with the state based power demand P, the cycle and total time (t) and the production volume n (Engelmann, 2009):

$$W = P * t_{cycle} * n + n\left(t_{total} - \left(n * t_{cycle}\right)\right) * P_{idle} \tag{6}$$

The equation reveals the energy demand when not producing (often equal to the constant power portion) in combination with the actual idle time, both can have significant impact on the overall energy consumption of a machine. Devoldere et al. conducted time studies for different types of production machines in industry cases, which underlined this relevance of “non-productive” energy consumption. This is specifically true if a major share of process-related inevitable handling and measuring activities is involved (e.g. in bending) (Devoldere et al., 2007; Devoldere et al., 2008). As a consequence, the energy consumption during non-production time should be reduced through organisational (e.g. influencing worker behaviour, batch processing) and technical (e.g. standby-mode) measures. Additionally, not only normative values but also state related energy consumption values should be considered when acquiring new production machines (Kuhrke et al., 2010).

2.4 Description of Selected Relevant Energy Flows in Manufacturing

The prior description revealed the relevance of different energy flows in manufacturing companies. Due to their relevance for this book, selected forms of energy will be presented in more detail. This will allow a deeper understanding of specific characteristics and provide the necessary theoretical background information for latter considerations.

2.4.1 Electricity

Electrical energy – or just electricity - is of major relevance in industrial practice due to some significant advantages in comparison with other energy carriers (Schufft, 2007; Ridder, 2003). Electricity,

- can basically be generated from any primary energy carrier
- has relatively few losses in distribution
- can be measured easily and also used for control purposes
- is completely convertible without any residues
- has a significant relevance in information and communication technologies (ICT)

However, three major disadvantages need to be stated as well (Schufft, 2007):

- Electricity is basically not storable.
- Electricity has to be generated in conversion processes with relatively low efficiencies.
- Extensive facilities are necessary for electricity generation and distribution.

As described above, electricity is being generated in power plants through using coal, gas, oil, nuclear power or renewable sources such as wind, solar power or hydropower. In the electric grid, alternating current is used because of the advantages in comparison with direct current. Distribution involves diverse current levels, which are obtained through different transformation steps. Figure 20 shows the structure of an electricity supply system with generation and distribution to consumers, which differ in terms of their connection due to their power demand. Manufacturing companies are typically connected to the high or mean voltage grid, in smaller cases (e.g. SME) low voltage might be sufficient as well.

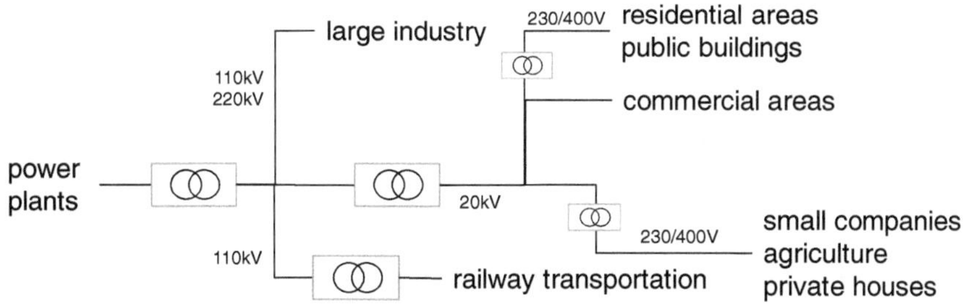

Fig. 20 General structure of electricity supply system (Schufft, 2007)

Pricing

The calculation of electricity costs in manufacturing companies is very complex and is influenced by diverse variables. Additionally there are typically no standard contracts but rather very specific conditions for each company as a result of individual negotiations between the supplier and customer (Specht, 2005). Based on analyses of different electricity supply contracts, this section just gives a brief summary of relevant cost aspects and calculation procedures. Three different main cost portions can be distinguished:

- costs for electrical work ($Electricity\ Costs_{work}$)
- costs for electrical power ($Electricity\ Costs_{power}$)
- fixed and variable standard costs ($Electricity\ Costs_{standard}$)

Therewith, the composition of total electricity costs ($Electricity\ Costs_{total}$) sums up to

$$Electricity\ Costs_{total} = Electricity\ Costs_{work} + Electricity\ Costs_{power} + Electricity\ Costs_{standard} \quad (7)$$

In the following, these cost portions will be described briefly. The base for electricity cost calculation is a time based electrical load curve as exemplarily shown in Figure 21. The measuring respectively relevant billing interval is typically 15 minutes – the measured electrical power demand (which be technically available with significantly higher sampling rate) is being averaged over this period (Müller, 2001). This results in 96 electrical power values for one day (24 hours).

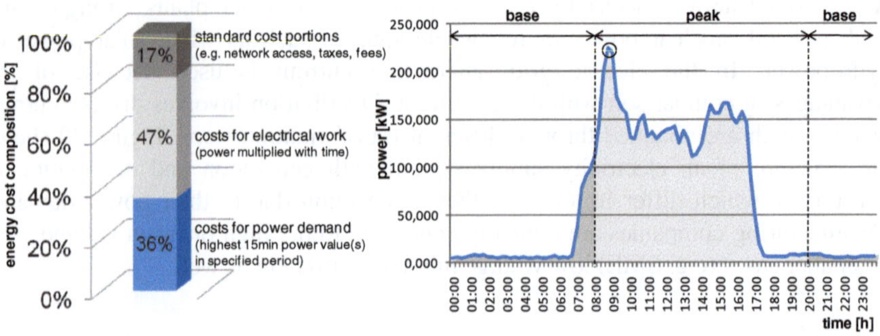

Fig. 21 Example of electricity cost composition and sample daily electrical load profile (own investigation based on actual data from company)

Costs for Electrical Work

Mathematically, the electrical work is the area beneath – and so the integral – of the electrical load curve. Based on equation (1) and a given price rate (€ per kWh) the costs of electrical work can be calculated with:

$$Electricity\ Costs_{work} = electrical\ work * price\ per\ kWh \quad (8)$$

In energy economics there is a differentiation between so called *peak* and *base* times. Against the background of typical market behaviour influenced by supply and demand, energy suppliers may charge a higher "price per kWh" during the day (peak, typically between 6am and 8pm) because of higher general demand. In base times (evening/night) this rate can be significantly lower because at this time energy suppliers want to utilise their power plants as well. However, as mentioned above, this is just a general framework and the specific solution for an individual company is the subject of individual negotiation – this might also mean that there is a single cost rate based on a mixed price model.

Costs for Electrical Power

In contrast to private household billing, industrial companies as relatively large consumers are typically also charged based on electrical power demand. Energy suppliers argue that power demand stresses the electric grid as a whole and guaranteeing capacity and reliability of the system involves significant effort. This cost portion can make up a very significant share on total electricity costs, especially in SME with volatile electricity consumption patterns. For the calculation of this cost portion the highest 15min-value of electrical power demand for a defined period of time (e.g. for a month) is being multiplied with the specific demand related price rate (€/kW)

$$Electricity\ Costs_{power} = highest\ power\ demand * price\ per\ kW \qquad (9)$$

Again, the actual calculation scheme and price rate is highly individual: for example, the highest electrical power value for a month could be the base of calculation. However, also cost calculations which take e.g. the three highest values over a whole year into account can be found in industrial practice.

Fixed and Variable Standard Costs

This electricity cost portion should subsume the manifold additional cost elements, which are being charged, connected either to the electrical work or just a fixed rate.

$$Electricity\ Costs_{standard} = \sum(electrical\ work * variable\ standard\ costs\ per\ kWh) + fixed\ standard\ costs \qquad (10)$$

Most of these issues are regulated by the government, such as taxes, fees for network access or country specific dues (e.g. in Germany for supporting the introduction of renewable energies) and cannot be negotiated. Additionally there are standard charges from the supplier for e.g. measurement or billing.

2.4.2 Compressed Air Generation

For industrial purposes compressed air offers some important advantages compared to e.g. hydraulic, electrical or mechanical systems (Bierbaum and Hütter, 2004). For instance, compressed air is:

- easy to store and transport (also possible over longer distances, no recirculation necessary)
- clean and dry medium
- safe to operate
- components of the system are often less expensive, more robust/simple, lighter and easier to maintain (e.g. no exchange of fluids necessary)
- fast medium
- overload normally causes no problem
- parameters infinitely variable

In industry, compressed air is typical used for e.g. clamping (e.g. fixation of parts in automated systems), transportation (e.g. bulk goods in pipes), pneumatic drives (e.g. automatic screwing), spraying (e.g. painting, blasting), blowing (e.g. glass industry, cleaning of parts), testing (e.g. pressure checks) or control tasks (Bierbaum and Hütter, 2004). The generation of compressed air is a conversion of electrical energy to mechanical energy. Physically, the compressed air system is determined by three main variables, temperature (T), pressure (p) and volume (V), which are thermodynamically directly connected with each other (Bierbaum and Hütter, 2004):

$$\frac{p*V}{T} = constant \tag{11}$$

Assuming one of the variables as being constant, the two others are changing proportionally to each other, e.g. (for constant temperature)

$$p_0 * V_0 = p_1 * V_1 = constant \tag{12}$$

The conversion process takes place in one or several compressors. Diverse kinds of compressors are available which differ in terms of functional principle (e.g. rotary and reciprocating compressors), energy consumption, compressed air volume rate etc, but also regarding control schemes (Bierbaum and Hütter, 2004; Ruppelt, 2003). A comprehensive European market study revealed that so-called screw compressors dominate the market (approx. 75% market share) because of their "reliability, simplicity and relatively low costs" (Radgen and Blaustein, 2001). The same study also gives an impression of the immense relevance of compressed air generation in terms of energy consumption. About 10% of total industrial electricity consumption is caused by generation of compressed air (which means 80 TWh or 55 million tons CO_2). The average compressor has a demand of 71 kW and runs 3,500 hours per year (Radgen and Blaustein, 2001). The compressor(s) are just one part of a whole compressed air system which consists of several other centrally controlled components for air supply and treatment (filter, dehumidification/drying), distribution (e.g. pipe network, puffer tanks) as well as air usage (e.g. production machines or manual devices such as blow guns) (Radgen and Blaustein, 2001; Ruppelt, 2003; Müller et al., 2009).

As one big disadvantage compressed air usage is often connected with very high system losses. Studies show that not even 10% of inserted energy ends up as usable mechanical energy at the end-use device (Figure 22). As a result compressed air is actually one of the most expensive forms of energy in industry

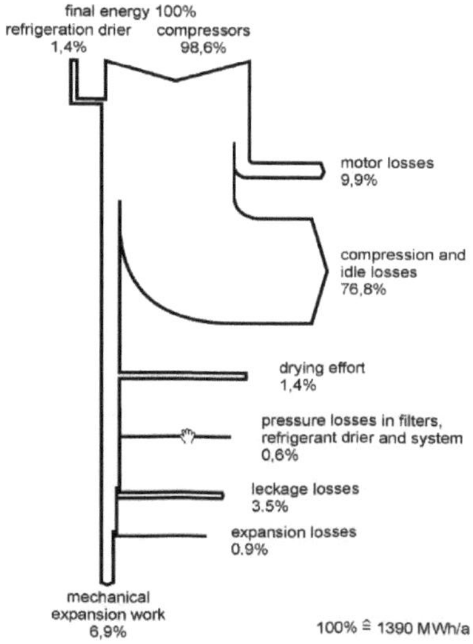

Fig. 22 Losses during the generation of compressed air depicted as Sankey-diagram (Gauchel, 2006)

(Gauchel, 2006; Gloor, 2000). As Figure 22 underlines, losses occur in all components of the system. The main amount of energy (>90%, Bierbaum and Hütter, 2004) gets lost as heat at the compressor (typically there are also leakage losses in the system). As orientation for evaluating the performance, Figure 23 shows typical specific compressor power demand for generating a compressed airflow of one m^3/min with respect to the nominal pressure of the system (typical range for regular industrial application) whereas the lower boundary is a loss-free isothermal compression (Gloor, 2000). Having in mind the broad application fields in industry underlines the huge additional energy demand caused by compressed air (e.g. a simple impact wrench needs approx. 0.2-1 m^3/min; a small sandblasting workplace alone needs at least 1 m^3/min – larger workplaces reach 3-5 m^3/min) (Bierbaum and Hütter, 2004; IES – Industrie Engineering Service GmbH, 2009) as well as the significant losses which are involved (even in a good case).

Literature mentions different areas for improvement of compressed air generation, e.g. (Radgen and Blaustein, 2001; Gloor, 2000; Saidur et al., 2010):

- improvement of drives in compressors
- optimal choice of type and dimensioning of compressor (over dimensioning as a problem)
- recuperating waste heat
- decreasing system volume (e.g. block pipes that are not in use)
- reducing nominal system air pressure
- network design (e.g. pipes, avoiding filtering/flow restrictions)

- optimising control
- avoiding air leaks
- improved air treatment

Studies reveal that these opportunities are not seized yet; saving potentials are estimated with 5-50% (average approx. 33%) for the next 15 years (Radgen and Blaustein, 2001).

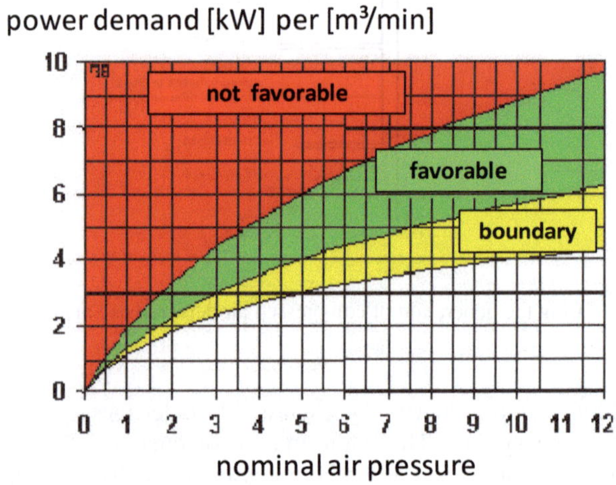

Fig. 23 Specific compressor power demand in kW for generating for one m³/min compressed air depending on nominal system pressure (Gloor, 2000)

2.4.3 Steam Generation

Steam is another important form of energy and widely used in different industries for diverse purposes besides its major role of generating electricity in power plants. For example, 37% of all fossil fuel burned in US industry are used to generate steam (Einstein et al., 2001). Typical producing industries with heavy steam usage are, for instance, the food, paper, pharmaceutical, chemical, textile, petroleum refining and primary metal industry (Spirax-Sarco Limited, 2005; Einstein et al., 2001). Steam is a relatively efficient way of generating and distributing heat which is needed in order to conduct certain production processes (e.g. sterilising, cleaning, drying, distillation, cooking) or provide space heating. Advantages of this form of energy are the high heat content and heat transfer capabilities, safety, cleanness and flexibility (Wagner, 2010; Spirax-Sarco Limited, 2005). The steam system consists of diverse components – water treatment, boiler, pipes, consumers, steam traps and condensate loop - which are depicted in Figure 24 (Spirax Sarco, 2006). Steam with predetermined pressure and temperature is generated through vaporising water through burning fuel (Rajan, 2008). The steam is distributed to its consumers where heat transfer takes

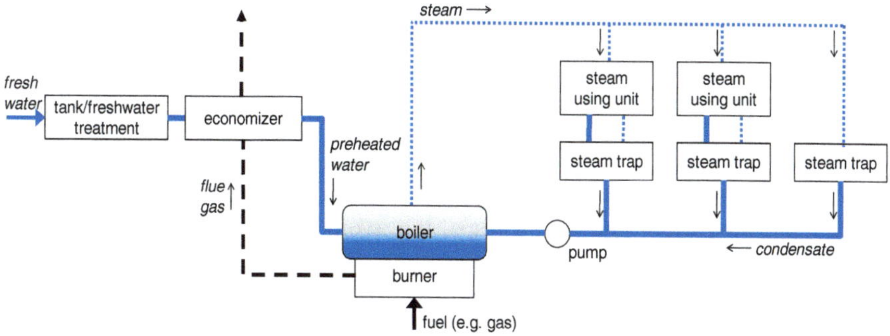

Fig. 24 System for steam generation and distribution (Spirax Sarco, 2006; Einstein et al., 2001)

place and the steam cools down to change the aggregate state back to liquid (condensation). This condensate is flowing back into the boiler to close the cycle and be vaporised again.

The steam generation process is the core element of the steam system. Within the steam raising unit, pre-treated (to avoid impurities, which may harm the system) water is brought to its boiling temperature before being vaporised in the boiler. In superheaters the steam can be further heated in order to achieve the designated temperature and pressure. The coherences between the determining variables (e.g. temperatures, pressure, enthalpy) in steam generation can be found in so called steam tables (e.g. Böge, 2009; Babcock and Wilcox Company, 2010). The necessary input energy for vaporisation comes from burning fossil fuels or biomass/waste used as fuel (Einstein et al., 2001) which results in flues emitted to the environment. Diverse types of steam raising units are available in industrial practice; an overview can be found in e.g. (Effenberger, 2000). They differ in terms of functional principle, possible volume rate and characteristics of the steam and efficiency which is defined as (in stationary operation and full load) (Recknagel et al., 2010):

$$\eta_B = \frac{boiler\ output}{heat\ input} = \frac{\dot{Q}_B}{\dot{Q}_C} = \frac{\dot{Q}_B}{FH} = \frac{boiler\ output}{fuel\ quantity * calorific\ value} \tag{13}$$

The boiler capacity is the decisive value to estimate the possible steam generation rate. The equation shows that the efficiency is determined by the supply rate and calorific value of the utilised fuel. The necessary amount of fuel ($\dot{m}_F$) with respect to the volume rate of steam ($\dot{m}_S$) as well as the specific enthalpy of steam and water (h_S, h_W) can be calculated with (Saacke, 2009):

$$\dot{m}_F\ or\ \dot{V}_F = \frac{\dot{m}_S * (h_S - h_W) * 100\%}{H * \eta_B} \tag{14}$$

Modern steam boilers achieve an efficiency of around 80-85%; main losses occur through heat loss in flue (approx. 75% of heat losses), radiation on surfaces and blow-down (Dalzell, 2000; Spirax-Sarco Limited, 2005). In addition to the boiler losses, the distribution system incorporates further losses of 5-30% (Dalzell, 2000). Studies reveal an economic potential of 18-20% for the improvement of energy consumption in steam systems (Einstein et al., 2001). In the case of the boiler, relevant leverages are an improved process control, the reduction of flue gas quantity and excess air (for combustion), improved insulation, boiler maintenance, blow-down steam and flue gas heat recovery (through economisers) as well as the usage of alternative fuels. For the heat distribution, improving and maintaining insulation and steam traps, the avoidance of leaks and circuitry of condensate are important issues (Einstein et al., 2001).

2.5 Energy Efficiency in Manufacturing

2.5.1 Definition

As mentioned above, efficiency is one of the three main strategies towards a Sustainable Development as promoted by e.g. the Brundtland Report (Brundtland Commission, 1987). In general, efficiency is defined as ratio of any output to the necessary input of a system (DIN 9000). For the case of energy it can be formulated as (EuP Directive 2005/32/EC, 2005):

$$Energy\ efficiency_{general} = \frac{output\ of\ performance, goods, service\ or\ energy}{input\ of\ energy} \tag{15}$$

This term is very generic and therefore "energy efficiency can mean different things at different times and in different places or circumstances" (European Commission, 2009). In technical systems (e.g. energy transformations in general in Chapter 2.2, see steam generation in Chapter 2.4.3) the term efficiency is often used thermodynamically as the ratio of output energy to input energy in order to assess the quality of energy conversions (Patterson, 1996).

$$Energy\ efficiency_{thermodynamical} = \frac{output\ energy}{input\ energy} \tag{16}$$

In the context of energy efficiency in manufacturing, a physical-thermodynamic meaning is typically applied, which means optimising the ratio of the production output (e.g. in terms of quantities with defined quality) to the total energy input (electricity, gas, oil) (Freeman et al., 1996; Quadriguasi et al., 2009).

$$Energy\ efficiency_{manufacturing\ system} = \frac{production\ output}{total\ energy\ input} \tag{17}$$

However, this is still a general and simplified equation. Both determining variables are not easy and generally definable and are subject of discussion in literature (Patterson, 1996). To correctly evaluate the energy efficiency of manufacturing systems the consistent definition of the system boundaries (e.g. machine, factory, country) as well as of the different possible input and output variables (respectively their units) is critical for application. This is specifically true since just one efficiency figure has a limited meaning – it is important to have a comparison of values in order to generate meaningful information about the state of a manufacturing system (Engelmann, 2009). For example, production output could be measured in terms of produced quantities (unit: pieces), masses (unit: kilograms/kg) or generated revenue (unit: Euro/€). Energy input can be expressed, for instance, by energetic value (e.g. in kWh), energy costs (unit: Euro/€) or environmental impact (e.g. induced CO_2 emissions) (Verfaillie and Bidwell, 2000).

The energy efficiency ratio should be as low as possible whereas physical boundaries set the lower limit. The following equation shows a more specific energy efficiency definition for the case of a production machine (or system of machines) powered by electricity for a certain period of time.

$$Energy\ efficiency_{per\ piece} = \frac{produced\ pieces}{electrical\ power\ demand * time} \tag{18}$$

One strategy towards energy efficiency is reducing energy consumption (through less power demand and/or less necessary time) while keeping output (at least) at the same level. However, the ratio clearly shows that increasing the output with no or proportionately less increase of the input energy is also a legitimate way for improvement.

2.5.2 Potentials and Fields of Action

Different studies reveal the significant improvement potential within industry. The study "Energy Efficiency in Manufacturing" carried out by Fraunhofer Gesellschaft underlines the relevance of production processes in single companies as well as on a global base and highlights the major potential of increased production process efficiency to optimise the environmental as well as economic performance (Fraunhofer IWU, 2008). The study "Energy Efficiency in Manufacturing: The Role of ICT" claims saving potentials of 10-40% in manufacturing and stresses the importance of ICT (Information and Communication Technologies) as an enabler for energy efficiency (European Commission, 2008). A comprehensive study carried out in Germany reveals similar significant potentials in the manufacturing industry regarding e.g. the efficient usage of energy (Seefeldt and Wünsch, 2007). Altogether, based on the available technology in 2002 and depending on the field of action, a saving potential of 10-30% on energy consumption was identified. From today´s perspective the potential is likely to be even higher. These rather research driven estimations of potentials are also confirmed through studies in industry. A broad survey on different companies revealed an average estimated potential of energy efficiency of 15% (Schröter et al., 2009).

Some general fields of action to increase the energy efficiency in manufacturing can be distinguished and will be briefly described in the following section.

Production Machine Level

Figure 25 shows the possible approaches to increase the energy efficiency on production machine layer. Relevant influencing variables are:

- The **design** of the machine which includes the general structure/functional principle (including definition of necessary energy inputs for different purposes), appropriate dimensioning and the selection of energy efficient components (e.g. efficient electric drives for spindle, pumps etc) (Kuhrke et al., 2010; Devoldere et al., 2007).

- Machine **control** offers saving opportunities through intelligent logic interaction of all machine elements (Dietmair and Verl, 2010; Cannata et al., 2010), e.g. intelligent shutdown (in idle mode).
- **Process parameters** can have significant influence, e.g. increasing the speed of processing often leads to higher energy consumption (Lanz et al., 2010). However, energy efficiency as key figure is important to consider since this increase does not have to be proportional. As mentioned in 2.3.3, the fixed energy consumption share is often relatively high; processing faster would therefore reduce the specific energy per part (Klocke et al., 2010; Boos and Kuhlmann, 2010).

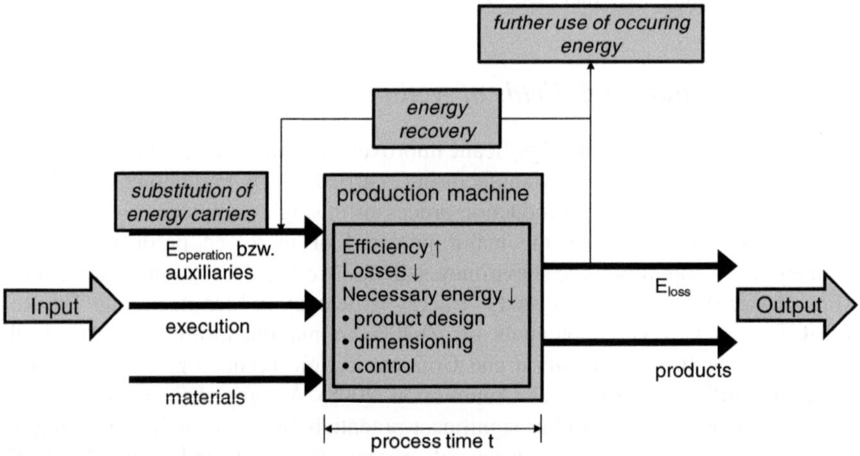

Fig. 25 Variables to influence the energy efficiency of production machines (Müller et al., 2009)

Production Planning and Control (PPC)

As mentioned before, the energy demand of a factory is determined by diverse consumers and adds up to a specific load profile of the different energy carriers. Figure 26 shows general possibilities to influence the cumulative load curve on factory layer (Gesellschaft Energietechnik, 1998). In this context, the avoidance of energy consumption peaks is an important aspect for manufacturing companies in order to avoid peak surcharges (e.g. electricity billing). Another option is the shifting of consumption to daytimes with less expensive energy price rates (e.g. base time at night) (Tönsing, 1996). These options are certainly focusing less on the decrease of energy consumption but increasing the energy cost efficiency. However, technical and environmental targets are also relevant. On the one hand,

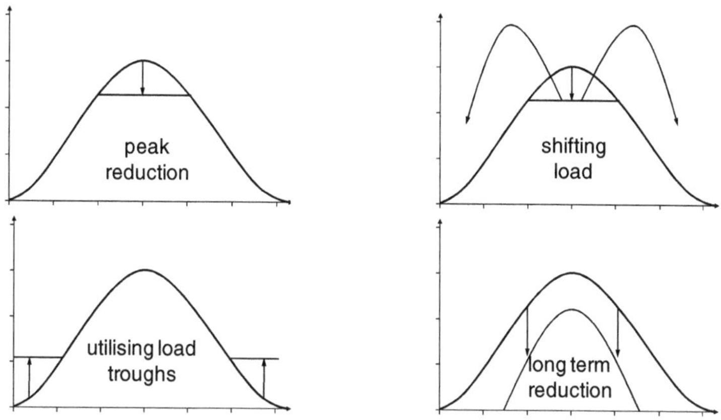

Fig. 26 Measures for influencing energy demand from factory perspective (Gesellschaft Energietechnik, 1998)

a balanced consumption behaviour supports continuous energy and media supply in favourable working points and therefore less capacity has to be provided. Altogether, this can result in less investment, less operating costs and also less energy consumption with its related environmental impact (e.g. dimensioning of compressed air system or electricity grid). Actual means for influencing the consumption behaviour are technical measures like energy/load management systems (e.g. Gesellschaft Energietechnik, 1998) or rather organisational measures like an energy-aware production planning and control (PPC, e.g. scheduling, lot sizes)(Schieferdecker, 2006; Bonneschky, 2002). Energy aware PPC in particular also allows energy consumption reduction through optimal utilisation of equipment and avoiding energy waste in idling machines. The impact of shifted production order sequences on energy demand is exemplarily shown in Figure 27 (Rager, 2008).

Supporting Processes Including Technical Building Services (TBS)

In the context of technical building services, the technical configuration of the equipment (e.g. dimensioning, materials, usage of energy efficient components, detail design of tubes etc), an efficient process control (e.g. high utilisation – avoid standby, continuous runs versus start-stop-operation, processes at favourable working point) as well as the avoidance of losses (e.g. leakage, lacking insulation) are main levers to improve e.g. energy efficiency. Simulation-based approaches are available to support the development of different measures (Chow, 1996; Andreassi et al., 2009; Wischhusen et al., 2003; Rebhan, 2002; Kircher et al., 2010). Some specific fields of action for the case of compressed air supply were already presented in Chapter 2.4.2, for steam in Chapter 2.4.3.

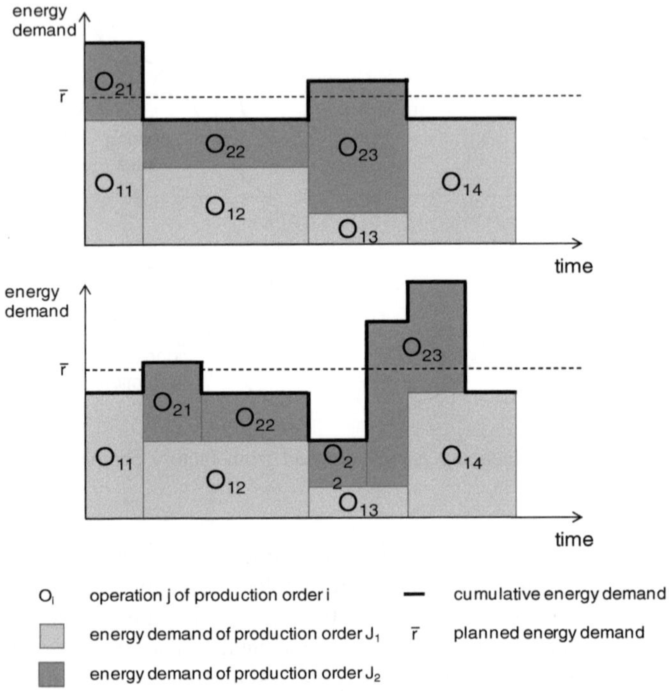

Fig. 27 Influence of PPC on energy demand (Rager, 2008)

Building Shell

In civil engineering the improvement of (energy) efficiency of the building itself is an important topic for many years, and therefore energy efficiency measurements (and labelling) are commonly used. Typical approaches focus on adequate design of building elements like insulation, roofing, walls, windows, slabs and foundations (Energy Star, 2010). The consideration of airflows within the building is a major element which is supported by diverse available simulation approaches (e.g. TRNSYS – software tool for simulation air/heat flows in buildings). A comprehensive overview of available tools can be found at (U.S. Department of Energy, 2010).

Chapter 3
Derivation of Requirements and Methodological Approach

The following section will first focus on the derivation of requirements for further support towards energy efficiency in manufacturing. This discussion is based on the theoretical background but also the industrial demand. These requirements serve as input for the later selection of the methodological solution approach being used.

3.1 Requirements from Industrial/Business Perspective

As underlined in the previous chapter; significant potentials to improve the energy efficiency in manufacturing companies are basically available. However, diverse studies based on comprehensive surveys also highlight that there are - specifically for small and medium enterprises (SME) - relevant obstacles which impede a broad implementation of promising measures despite the general willingness of many companies (Sorrell et al., 2000; Thollander, 2009; Thamling et al., 2010; Schmid, 2008). These issues should be consciously addressed and serve as requirements for a new solution approach. Based on diverse studies/references the following main obstacles (O) can be identified (Sorrell et al., 2000; Thollander, 2009; Gesellschaft Energietechnik, 1998; Beyene, 2005; Kaiser and Starzer, 1999; Offner, 2001; Schmid, 2008; Brüggemann, 2005; Thamling et al., 2010; Stern, 1984; Jaffe and Stavins, 1994).

- **O1:** Due to necessary investment for e.g. new technologies and the relatively low level of energy prices, energy efficiency measures may just sum up to **small cost savings with long amortisation time** - which make those measures not attractive to implement. Measures are often too selective and restricted to certain areas – larger approaches with more leverage are missing.
- **O2:** Companies often face **lacking access to necessary capital for implementing energy efficiency measures**. Capital is a scarce resource – if available, it is typically needed for investment in other areas of production.
- **O3:** There are **no resources for energy efficiency measures** – personnel and available time are normally allocated to other tasks in order to keep value creation running.

S. Thiede: Energy Efficiency in Manufacturing Systems, SPLCEM, pp. 35–49.
springerlink.com

- **O4: Lacking organisational responsibility** impedes a systematic consideration of energy consumption and efficiency.
- **O5:** There is very **restricted transparency regarding the energy consumption** in the company. Due to the lack of data, main consumers are rarely known and systematic improvement not possible. Additionally, the topic as a whole is **too complex** or assumed to be too complex.
- **O6:** Companies **lack the necessary detailed knowledge of energy efficiency** measures including an overview over available technologies.
- **O7:** Altogether, production objectives have often the highest priority whereas energy consumption is of lower priority. There is a **concern about potential negative influences on production** that have to be addressed.
- **O8:** While diverse means for improving the energy efficiency are identified and described in literature, not every measure may be transferable and suitable for the specific case of a company. Hence, **transferability respectively verifiability is an important issue** which would allow to estimate the impact of certain measures for the actual case.

Some obstacles are more an issue of politics (e.g. O2 is addressed through support programs, incentives towards energy efficiency measures) and also market behaviour (e.g. O1 – higher energy prices will automatically drive implementation). Evidently, keeping up value and creating production is of main interest. In this context, strategic and operative production management plays a key role as important interface to energy efficiency related concepts. Besides the determination of production activities to generate valuable outcome, production management also heavily influences the energy consumption of the factory system when planning (e.g. structuring production, selection and dimensioning of technical equipment) and operating (e.g. scheduling) the facility. However, studies underline that – compared to other variables - energy efficiency is still a minor relevant target objective even though specifically system planning has significant influence on energy consumption behaviour in later use phases (Engelmann, 2009).

Against the background of the mentioned obstacles and the relevance of production management, the following requirements (R) can be derived in order to support the implementation of energy efficiency measures in industry:

- **R1:** Considering energy efficiency should not be an additional burden in daily work. **No add-on but an integrative approach** is needed which delivers simultaneous perspectives on all time, cost, quality and energy related objectives. This avoids too much additional effort, supports thinking in cause and effect chains, enables the identification of impacts on production and automatically draws attention to the topic in daily work (addresses O3, O4, O5 and O7).
- **R2:** “Complex tasks, simple tools” – considering energy flows in manufacturing is very complex. However, while keeping this complexity in the background this task must be broken down to **simple, controllable solutions.** This enables operation in practice without too much effort while coping with the complexity of the issue (addresses O3 and O5).

- **R3:** There should be appropriate **means for easy identification of main consumption drivers**. This allows concentrating on promising fields of action with large effects, serves as filter for target oriented knowledge acquisition (e.g. deep analysis of available technology) and reduces unnecessary efforts (addresses O1, O3, O5 and O6).
- **R4:** It is necessary to be able to **evaluate energy efficiency measures on different levels within one solution**, including their interdependencies. For example, it should be possible to assess and compare both the impact of smaller improvements (e.g. installing efficient drives in one machine) but also of more complex solutions (e.g. energy aware production control). This allows prioritising possible measures (for sequencing implementation) and fosters systematic improvement (addresses O1, partly O2, O5 and O7, O8).
- **R5:** Finally, the **realistic and reliable calculation of key figures** is a major issue, because they serve as base for implementation decisions. This includes the realistic calculation of actual (energy) costs and the relevant production performance criteria (e.g. output, throughput time, utilisation) as well as the necessary consideration of uncertainty and risk (addresses O1, O5 and O7).

3.2 Requirements from Scientific/Technical Perspective

In addition to requirements from industry as potential users, there are further aspects from scientific respectively technical perspective which have to be addressed by an appropriate solution approach. Section 2 gave an overview on aspects that have to be considered in context of energy efficiency in manufacturing. It revealed the enormous complexity of the topic, which is mainly determined by (Schmid, 2008):

- the multitude of elements (each single light bulb or electric drive influences the energy consumption)
- the diversity of elements (entirely different technologies and disciplines involved)
- quantity and diversity of element interactions in context of energy consumption
- the time based volatility of energy consumption that is based on seasons, intensity of usage etc

A realistic and goal-oriented identification of improvement potentials and actual measures without the danger of problem shifting, demands a comprehensive system-oriented view. Against this background, further critical requirements (R) towards a holistic comprehension of the factory system can be derived (first published in Herrmann et al., 2010a).

- **R6 - Extended process comprehension:** All relevant input and output flows of production processes must be explicitly considered in order to avoid focusing on less relevant issues (while neglecting major challenges) and local optimisation with problem shifting. This includes all energetic (e.g. compressed air, electrical power, waste heat) and material (e.g. auxiliary materials as cooling lubricants) flows, which lead either directly or indirectly to additional energy and/or resource consumption. Against this background Figure 28 shows an extended process model, which integrates previously separately considered ecological and economic process perspectives (Schultz, 2002; Herrmann and Thiede, 2009a).

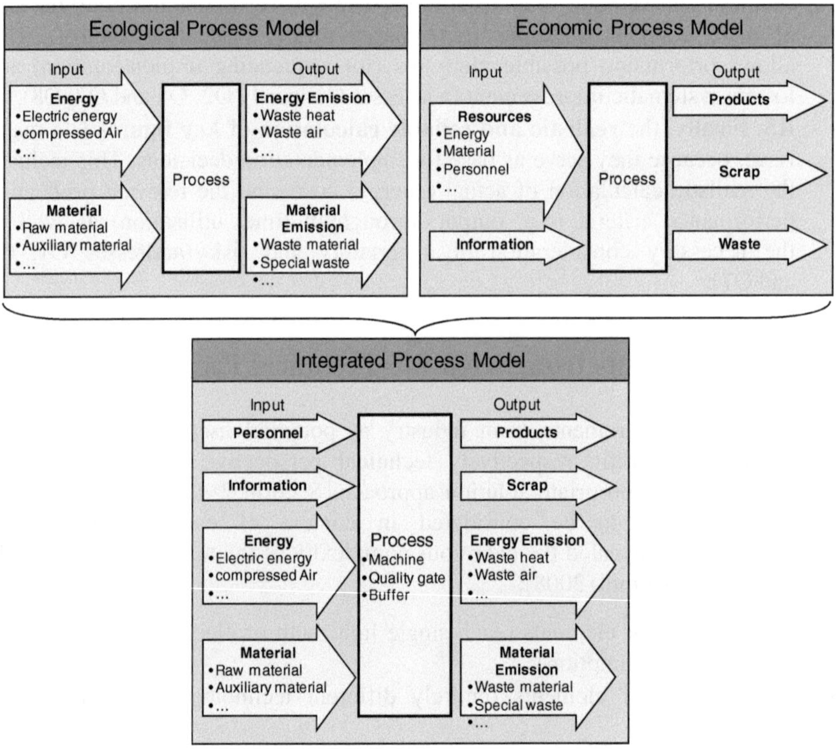

Fig. 28 Integrated process model (based on Schultz, 2002)

- **R7 - Holistic system definition of the factory:** Manifold interdependencies between the constituting elements of a factory also demand for an enhanced comprehension of the factory system as a whole. This system can be divided into the three sub-systems, namely production (machines and employees, coordinated by production planning and control), technical building services (TBS) and building shell. Altogether, this results in a complex control system with dynamic interdependencies between different internal and external influencing variables (Figure 29, Hesselbach et al., 2008b).

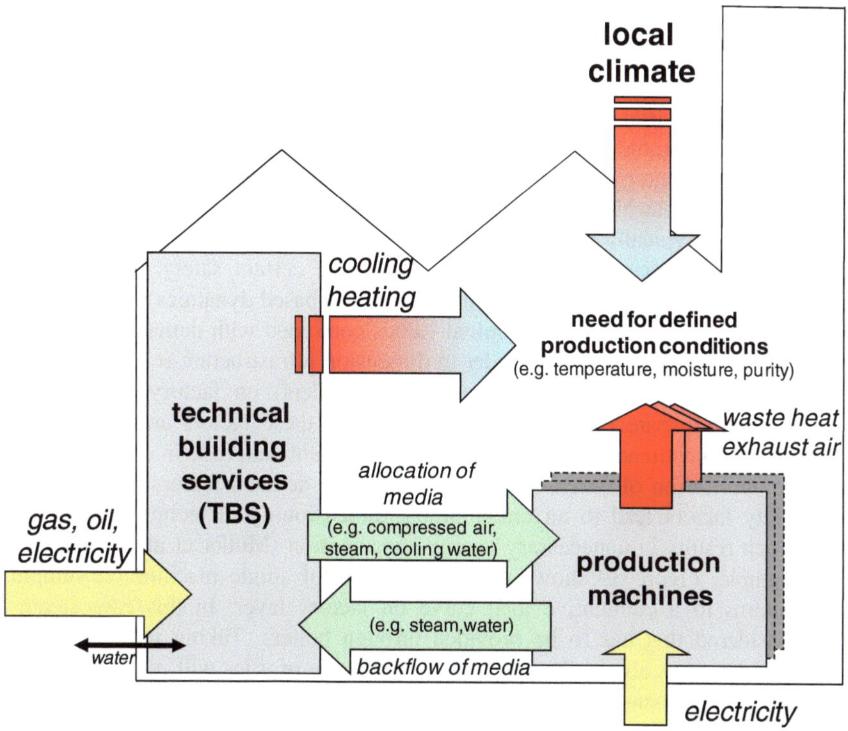

Fig. 29 Holistic definition of factory (own illustration, first presented in Hesselbach et al., 2008b)

As shown in Figure 29, one major task of technical building services is to ensure the needed production conditions in terms of temperature, moisture and purity through cooling / heating and conditioning of the air. The essential influencing variables are the local climate at the production site (seasonal influences) and the exhaust air and the waste heat that is primarily emitted by production machines but also by other production factors like transportation equipment or even personnel. Furthermore, production machines need energy (mostly electricity) and diverse media like compressed air, steam or cooling water to fulfil their designated processes. Technical building services are also responsible for the supply of these essential media whereas this involves their generation, their circuitry as well as the required conditioning (e.g. temperatures, pressures, purity). Altogether an evaluation of a factory's energy consumption must consider all non-regenerative energy flows that are externally supplied (e.g. electrical power, oil, gas) for running technical equipment for both production and TBS. Evidently, these sub-systems strongly interact with each other - they cannot be seen independent from each other when pursuing a holistic approach towards energy efficiency in manufacturing.

- **R8 - Dynamics of consumption and emission behaviour on machine and factory level:** As shown before, all relevant input and output flows are typically not static values but highly dynamic depending on the operating conditions of the processes and the machines. These profiles add up to cumulative load profiles on the factory level. In the end these dynamic cumulative load profiles (e.g. process heat demand, compressed air demand, heat flow into the factory building, electrical power demand) are decisive for design and control of the technical equipment (e.g. dimensioning of compressed air system) as well as for billing (e.g. energy supplier). However, the specific consumption and emission patterns of machines and the system as a whole are rarely available. For machines, typically just single nominal values are available which describe a maximum demand and inherently incorporate certain safety factors. Hence, these values neither reflect magnitude nor state based dynamics of consumption sufficiently. Even though, nominal values combined with demand/simultaneity factors are often applied in order to dimension infrastructure (e.g. electric grid, compressed air system, air conditioning, boilers) on factory level. Due to lacking transparency of consumption patterns these factors are typically also just rough estimations. Altogether, the insufficient data base in connection with the application of unrealistic (while unknown) demand factors plus additional safety factors lead to an unfavourable dimensioning of technical equipment, which results in unnecessary energy consumption (Müller et al., 2009). As an example, Figure 30 shows the superposition of single machine consumption patterns to a cumulative load curve on factory layer. In this case steam is considered that has to be provided through boilers. Taking into account the considerations above, the alternative cumulative profiles will result in different necessary dimensioning of the boilers as well as different energy costs (e.g. through consumption peaks and different idle consumption).

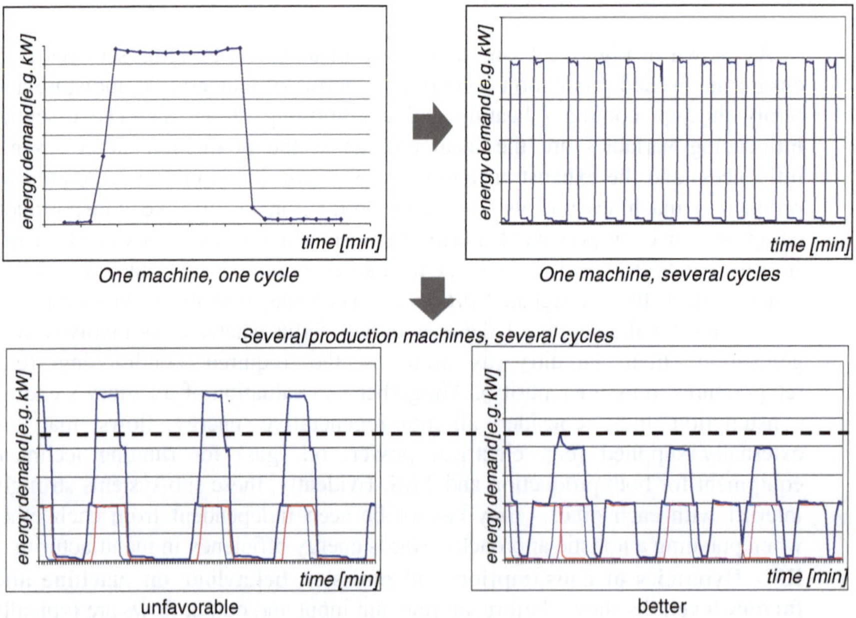

Fig. 30 Steam demand of one and several machines

- **R9 - Thinking in process chains:** Final products are usually not the result of single production processes, but are rather manufactured in several steps on different production lines in the sense of production process chains. Against the background of energy and resource efficiency, the process chain has to be regarded and evaluated as a whole, as it may involve further potentials (e.g. combination of processes). Moreover, problem shifting might occur while improving measures in one process can possibly lead to worse performance of others.
- **R10 - Life-cycle-oriented perspective:** Analogous to the thinking in process chains, all life phases of products (this includes also all the technical equipment within the factory itself) have to be considered when it comes to deriving measures concerning energy efficiency. Thus, the decisive factor for increasing the energy efficiency of a machine tool, for instance, is less the improvement of single parameters of a specific process than rather the development of the machine itself. Moreover, the choice of specific processes (e.g. joining techniques) has direct effects on the use and disposal phase which could lead to increased efforts in those phases. The life-cycle-oriented perspective also stresses the importance of energy aware planning of production facilities as the energy consumption behaviour is mainly determined in the planning phase (Herrmann, 2009; Herrmann et al., 2007a).
- **R11 - Consideration of all sustainability dimensions and integrated evaluation:** In order to deduce advantageous solutions, several relevant target dimensions have to be considered simultaneously. Going even further than requirement R5, besides a realistic economic (on the basis of a suitable cost model which integrates real contract conditions) and technical evaluation (e.g. effects on product quality), this includes an ecological evaluation (with a correct balance of the different input and output parameters, e.g. environmental effects of electricity and gas consumption). Possible conflicts of goals must be disclosed and decision support to solve them has to be offered (Herrmann, 2009).

3.3 Research and Methodological Approach

As short summary of the requirements derived above the holistic solution approach should be an **easy to use** solution for decision support in **strategic and operative production management** under **simultaneous consideration** of

- all relevant energy flows and connected dynamics,
- energy driven objectives with conventional time, quality and cost based targets,
- realistic calculation of energy induced costs and environmental impact,
- means towards energy efficiency on both single machine as well as factory system level.

As described before, single machine considerations are necessary but not sufficient for a holistic perspective on energy efficiency in manufacturing. Hence, the estimation of time based consumption patterns on a factory level is of major importance since this is the decisive base for

- dimensioning infrastructure respectively supporting processes like technical building services,
- realistic evaluation of energy costs, environmental impact and technical performance,
- derivation and assessment of energy efficiency measures going beyond single machine improvement.

For considering factory system consumption profiles four alternative methodological approaches could be pursuit: static calculations, fuzzy logic, artificial neural networks and simulation.

Static Calculations

Through combining available single machines consumption information and production data, the energy demand of a factory can be calculated statically using standard office software like MS Excel®. Figure 31 shows an analysis for the electricity consumption of a SME company. The electricity consumption of each major consumer was measured and – ex-post - put together with recorded production data (e.g. production program, machine runtime). The figure shows the good approximation in comparison with the electricity consumption as measured by the energy supplier (for the considered month the variance is even below 1%).

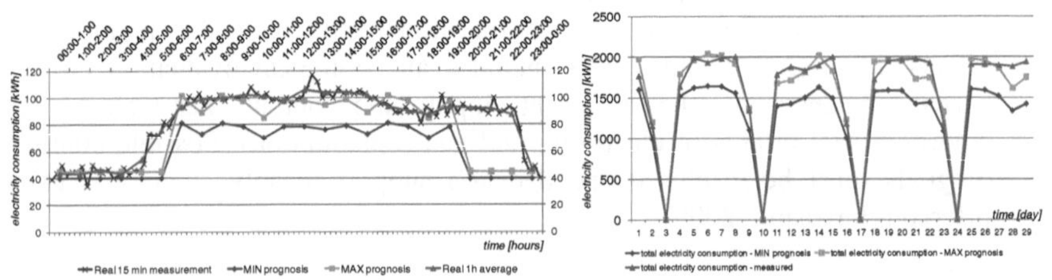

Fig. 31 Static ex-post calculation of electricity consumption and comparison to actual values (left: daily profile, right: monthly values)

The method increases transparency regarding energy flows and allows a good first estimation of consumption patterns in factory as well as a simplified evaluation of selected efficiency measures. However, while being an ex-post consideration, it obviously relies on the existence of detailed production data for the considered period of time. Ex-ante analyses based on planned production program are possible but would hardly achieve the necessary accuracy since uncertainties and interdependencies cannot be considered. This is specifically true for more complex cases with more consumers and energy flows (and more complex efficiency measures), which would also result in significant higher demand for calculation.

Fuzzy Logic

Lau et al. presented a fuzzy logic approach (Yager and Zadeh, 1992) to forecast energy consumption change in a manufacturing company (Lau et al., 2007). It aims at predicting the amount of energy consumption change compared to a reference energy consumption level that is based on selected production input variables. This shall help to e.g. consider whether the necessary power can be supplied. In their case study, Lau et al. applied their methodology to a clothing manufacturing plant. Based on the input variables (recorded for a certain period of time) daily total mass of finished products, total labour hours of operators in the plant in one day and the total running time of equipment in the plant they predict an additional demand on energy of 2.64% for this period compared to the reference consumption (Lau et al., 2007).

The fuzzy logic approach is based on a black-box-oriented perspective of the manufacturing system without detailed consideration of actual single consumers and their interdependencies. Furthermore, the method can provide the energy demand but not the necessary time based load profiles. These factors combined with the specific expertise needed for application restrict the applicability in context of the aforementioned requirements.

Artificial Neural Networks (ANN)

Some authors propose the usage of Artificial Neural Networks (ANN) to predict the consumption profile of a company (Fiedler et al., 2007; Lang and Hesselbach, 2009; Hufendiek and Kaltschmitt, 1998). Measured consumption data is used to train the network and the impact of selected influencing parameters can also be considered. A well trained ANN allows a good estimation of the time based consumption profile of e.g. electricity or gas (Fiedler et al., 2007). However, similar to the fuzzy logic approach the manufacturing system is considered as black-box being determined by selected influencing variables. It is not possible to break consumption down to single consumers. Additionally, ANN is a relatively complex method, which requires expertise and time for network setup and training.

Manufacturing System Simulation

Simulation is an established method within the planning phase of manufacturing systems and diverse mature software solutions (e.g. Plant Simulation™, Delmia™) are commercially available. However, energy related aspects are so far not being considered in these tools. In research, first approaches can be found that aim at augmenting material flow simulation with energy consumption data. Furthermore, detailed simulation models for e.g. technical building services or the building itself are available. In general these approaches are basically enabled to estimate the cumulative energy demand while keeping the necessary level of detail regarding cause and effect relations. Whereas being able to consider interdependencies and uncertainties (e.g. waiting times, failures) also a relatively reliable ex-ante prediction of energy demand might be possible. However, also simulation is more an expert tool and involves costs (licence) and time for model setup, validation/verification and simulation studies.

Table 3 shows a simplified evaluation of the four alternative methodological approaches based on a ranking with respect to the possible fulfilment of the identified requirements. A high ranking does not mean that all requirements are already fulfilled in those approaches – it simply indicates that it would be possible to address them properly from a methodological perspective (and compared to the alternatives). Manufacturing simulation clearly scores highest in almost all categories. It would allow addressing all requirements appropriately. Static calculation scores second but has major disadvantages when it comes to more complex energy efficiency analyses. ANN offers some advantages; at least once a well-trained network is available. It does however not allow deeper looks into the factory, which limits fields of application. The applicability of fuzzy logic is even more limited since not even a cumulative load profile can be predicted. Besides that it is a rather complex approach with limited usability in daily industrial practice.

Table 3 Evaluation of general methodological approaches based on identified requirements (ranking for each requirement from first to fourth place)

	Static calculation	Fuzzy Logic	Artificial Neural Networks (ANN)	Manufact. System Simulation
R 1: no add-on, integrative approach	2	4	3	1
R2: simple, controllable solution	1	4	3	2
R3: identification of main consumption drivers.	2	3	3	1
R4: evaluation of energy efficiency measures on different levels	2	3	3	1
R5: realistic and reliable calculation of key figures	4	2	2	1
R6 - extended process comprehension:	2	3	3	1
R7 - holistic system definition of the factory	2	4	3	1
R8 - dynamics of consumption behaviour	2	4	2	1
R9 - Thinking in process chains	2	3	3	1
R10 - Life-cycle-oriented perspective	1	3	3	1
R11 – integrated consideration of all sustainability dimensions	2	4	2	1
Mean Average Ranking	**2.0**	**3.4**	**2.7**	**1.1**

As a consequence, manufacturing system simulation is selected as methodological approach for the later concept. In the following some theoretical background will be provided.

3.4 Simulation Background

Simulation in general can be defined as "imitation of the operation of a real-world process or system over time" (Banks, 2010, similar definitions can be found in e.g. VDI, 2007). In this context, a system is defined to "be a collection of entities, e.g. people or machines that act and interact together toward the accomplishment of some logical end" (Law, 2007 referring to Schmidt, 1970). Thereby, a system is determined by its "structure (the inner constitution of a system) and behaviour (its outer manifestation)" (Zeigler et al., 2000). The behaviour of a system determines the relationship between its inputs and outputs. The internal structure includes the system's "states and state transition mechanisms" (Zeigler et al., 2000). For simulation, an appropriate model is necessary, which is "a representation of a system for the purpose of studying that system" (Banks, 2010). In contrast to a physical model, a mathematical model is meant in this context (Law, 2007).

Simulation is an established method for years in research and industry. Several examples of diverse applications can be found in (Engelhardt-Nowitzki et al., 2008; Bayer, 2003; Krenn, 2007); whereas a comprehensive overview over recent simulation applications provides (Jahangirian et al., 2010). Simulation offers some important advantages (summary based on Banks, 2010; Pegden et al., 1995; Law, 2007; Chung, 2004):

- Testing alternatives (e.g. policies, procedures, system design and control strategies) without disturbing the real system
- Testing of new system elements without necessity to acquire them (e.g. transportation systems, layout)
- Time can be compressed or expanded, which allows detailed analysis of interactions and/or long-time studies
- Simulation can provide deep understanding of the system with respect to its cause and effect relations (interaction and importance of variables). It "can help in understanding how the system operates rather than how individuals think the system operates" (Banks, 2010) and can make it visible to all stakeholders.
- A detailed analysis of e.g. hypotheses, certain phenomena or new design concepts is possible. This includes bottleneck analysis and answering "What if?" questions in order to improve the system.

However, simulation also involves some drawbacks which have to be addressed carefully when conducting studies (e.g. Banks, 2010; Chung, 2004).

- Simulation requires expertise and each model is influenced by the individual who created it.
- Simulation results can be difficult to interpret and their quality directly depends on the quality of input variables. The selection of appropriate performance criteria is critical and evaluation may involve statistical analyses.
- Simulation requires significant effort in terms of time and costs. For each case it should be carefully considered whether other (analytical) solutions might be possible or preferable.

Simulation approaches can be distinguished with respect to different criteria (e.g. Law, 2007). For example, **static** simulation models representing a system at a certain point of time (e.g. Monte-Carlo Simulation) whereas **dynamic** models are considering the evolution over a period of time. **Deterministic** models do not involve stochastic effects and lead to a distinct, repeatable result. In contrast to that **probabilistic** models involve one or several random variables (described through certain distribution function) and different runs will have different results. Therefore statistical treatment is an important issue.

Furthermore, a major distinction can be made between **discrete** and **continuous** simulation models. In discrete systems "state variable(s) change only at a discrete set of points of time" (Banks, 2010) whereas in continuous systems "the state variable(s) change continuously over time" (Banks, 2010; Cassandras and Lafortune, 2008). This important differentiation is shown in Figure 32. Nowadays, also so called **combined discrete-continuous or hybrid-simulation approaches** exist, which are able to combine both perspectives. Thereby, discrete events may change the state of a continuous variable or (vice versa) a continuous variable which achieves a threshold triggers a discrete event (Pritsker, 1995).

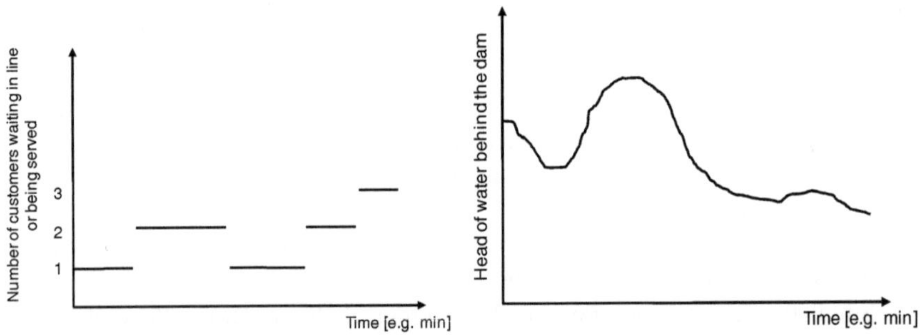

Fig. 32 Example of discrete (left) and continuous (right) state variable (Banks, 2010)

Based on these classifications four main simulation paradigms can be distinguished, which are displayed in Figure 33. Due to their different fields of application and logic, these paradigms were used by different research disciplines and implemented using different software tools (Borshchev and Filippov, 2004).

- **Dynamic Systems Simulation** is typically based on state variables and algebraic equations (often depicted through block diagrams), which describe the behaviour of physical systems. The standard tool for this application is MATLAB (www.mathworks.com/products/matlab).
- **System Dynamics Simulation** was invented by J.W. Forrester in the 1950's and describes the system with balancing or reinforcing feedback loops modelled as stock and flows (Forrester, 1969). It typically addresses problems on a higher level of abstraction such us whole ecological or economic systems (e.g. market behaviour, Sterman, 2000). An established tool for system dynamics modelling and simulation is Vensim (www.vensim.com).

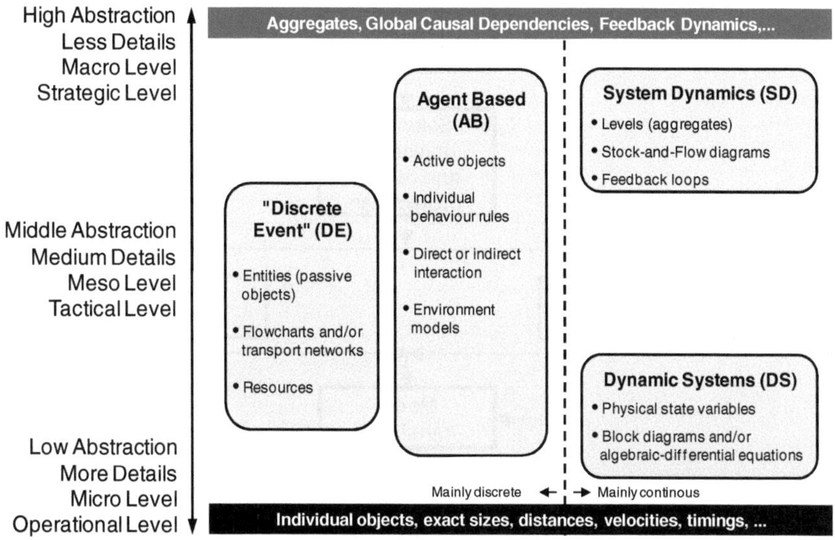

Fig. 33 Overview simulation paradigms (Borshchev and Filippov, 2004)

- **Discrete Event Simulation** is widely used in research and industrial practice for diverse purposes. Typically, passive entities (which represent e.g. products, people, messages) travel through a system (e.g. blocks in a flowchart) where they trigger certain actions and being e.g. delayed, processed, or combined (Borshchev and Filippov, 2004). An established application is the simulation of material flows in manufacturing systems. Therefore diverse specialised software tools such as Plant Simulation™ (www.plant-simulation.com) or Delmia™ (www.3ds.com/products/delmia) are available.
- As relatively new paradigm **Agent Based Simulation** focuses on decentralised modelling of individual object behaviour in a defined environment via state charts (Borshchev and Filippov, 2004). Each object acts individually based on his inherent logic and interacts with other objects. Altogether this leads to the behaviour/state of the considered system as a whole (e.g. transportation, pedestrians).

Some more recent software tools allow the integrative application of diverse paradigms, e.g. for combined discrete-continuous simulation models. Examples are Arena (www.arenasimulation.com, Kelton et al., 2010), Extend (www.extendsim.com) or specifically AnyLogic (www.xjtek.com/anylogic, Borshchev and Filippov, 2004).

Independent from the underlying paradigm some typical phases for conducting simulation studies can be distinguished (Banks, 2010; Law, 2007; Wenzel et al., 2008; VDI, 2007). An example is shown in Figure 34.

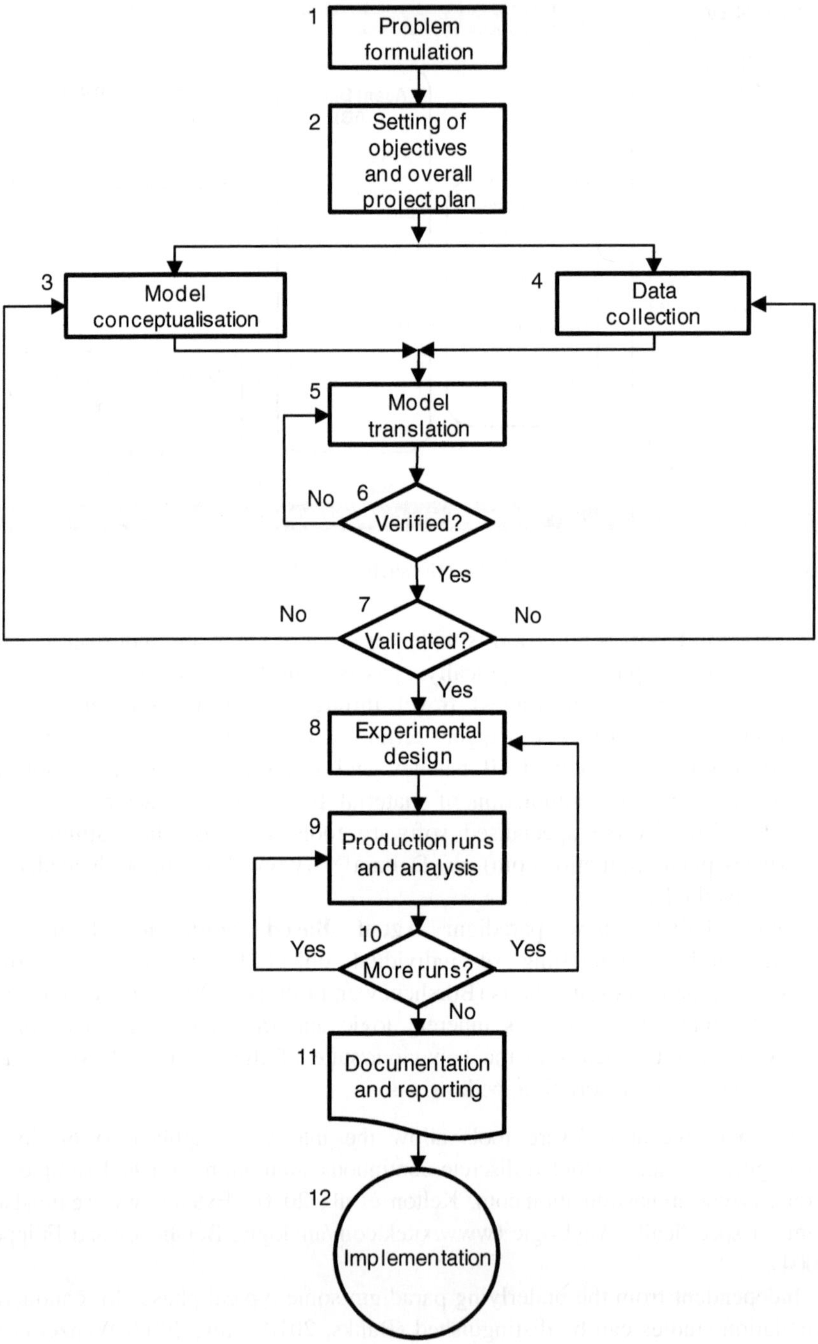

Fig. 34 Steps in a simulation study (Banks, 2010)

Verification and validation play an important role for reliable and realistic simulation results. Verification stands for the formal proof that the model itself is correct and its behaviour is consistent with other models (e.g. description model) (Rabe et al., 2008). Validation is the assessment whether "the model behaves with satisfactory accuracy consistent with the study objectives" (Banks, 2010). Practically, this could be the comparison of a simulation run with the comparable situation of the real system whereas the results should be consistent. A collection of available methods for validation and verification (differing in terms of applicability and subjectivity) can be found in (Rabe et al., 2008); an overview is shown in Figure 35.

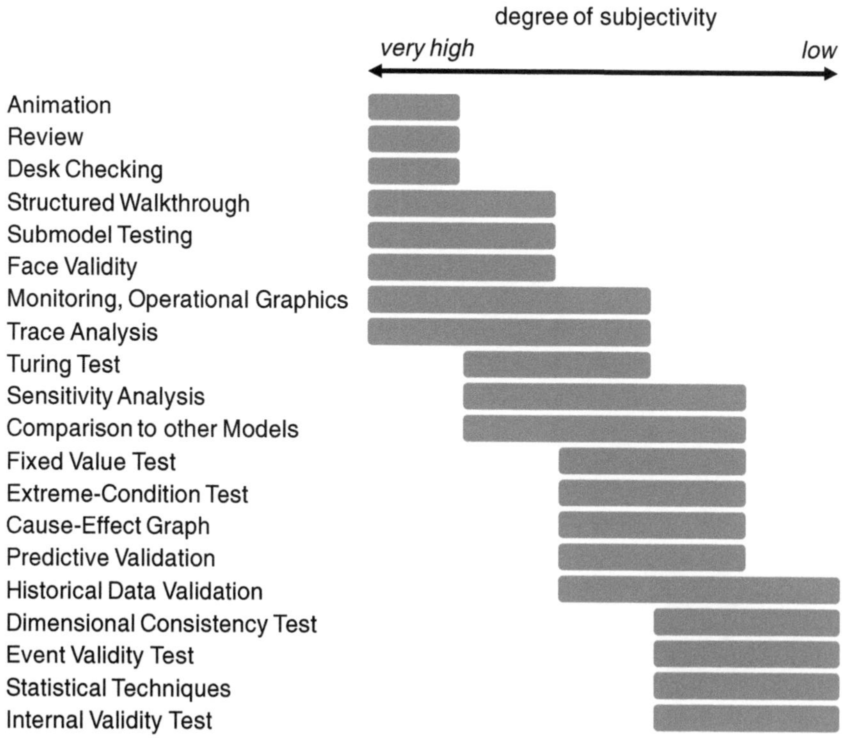

Fig. 35 Techniques for Verification and Validation and their subjectivity (Rabe et al., 2008)

Chapter 4
State of Research

The latter section revealed that simulation techniques are necessary to be able to realistically consider energy efficiency in manufacturing systems. There is a need to combine manufacturing system respectively factory level oriented simulation approaches with realistic representation of energy flows. Against this background the following section analyses and evaluates existing research approaches in this area with specific criteria derived from aforementioned requirements. Based on this detailed consideration further research demand will be deduced.

4.1 Background for Selection and Evaluation of Existing Approaches

Procedure and Limitations of Analysis

For the matter of this analysis a deep review of relevant books and research papers in the field of (e.g. manufacturing) engineering but also adjacent disciplines (e.g. operations research, computer science) was conducted. Taking into account the reasoning of the chapters before, necessary limitations regarding the selection of research approaches can be derived directly:

- While being an appropriate and necessary method to capture the dynamics of the factory system, only research approaches involving the **simulation in (discrete) manufacturing** are being considered in detail. To be more specific, considered approaches should (at least partly) involve **discrete event simulation** while this is the most established paradigm for manufacturing system simulation. This excludes pure mathematical (optimisation) models (e.g. Larsson and Dahl, 2003; Ferretti et al., 2008) and/or specific approaches for industries which are not in focus of this book (e.g. agriculture Neumann, 1985, chemical/process engineering Mark et al., 2009).
- Numerous approaches for different fields of application can be found which are focusing on the simulation of single processes or machines (e.g. FEM – finite element simulation – of machining processes). However, **multi-machine manufacturing systems** are considered here.
- While being of major interest, relevant **energy related flows need to be considered at all and with a certain level of detail**. Just naming a cumulative energy demand on a high level of aggregation (e.g. for calculation

S. Thiede: Energy Efficiency in Manufacturing Systems, SPLCEM, pp. 51–88.
springerlink.com

of overhead costs) is not sufficient. Energy consumption should be broken down to at least defined production areas if not individual machines with appropriate time resolution.

- A **direct relation to and certain focus on manufacturing systems** is necessary. This excludes research work being done to dynamise LCA (Life Cycle Assessment) with a significantly broader scope. These approaches certainly include one or even several production process(es) but specific interactions (e.g. queuing, machine behaviour) are rarely considered (examples are Gäbel and Tillman, 2005; Östergren et al., 2008). Additionally this limitation excludes research that exclusively focuses on the simulation of technical building services without direct relation to the production environment (e.g. Kircher et al., 2010; Wischhusen et al., 2003; Andreassi, 2009).

Definition of Criteria

Against the previously discussed theoretical background and connected requirements for the pursued solution approach, manifold specific criteria were derived, which will serve as base for evaluating relevant research approaches (Figure 36).

These criteria are structured in four main areas and described in the following oriented towards an idealistic degree of compliance.

Consideration of Energy and Resource Flows: Referring to the described theoretical background, an ideal approach has to consider all relevant energy and resource flows. This includes all forms of energy including externally supplied gas, oil or electricity as well as internal energy flows like compressed air, cooling water or heat. Additionally, raw or auxiliary material consumption might be of interest as it has direct and indirect influence on energy and resource efficiency of the manufacturing system. Not all flows are important in every case. However, the relevance of considered and ignored flows and possible limitations should be clearly pointed out (criterion *completeness of energy and resource flows*). Energy and resource consumption is not static but depending on machines respectively process states and time. This dynamic consumption behaviour has to be considered with sufficient accuracy. This is also true for the cumulative load profiles on manufacturing system respectively factory level, which are important to e.g. dimension and control technical building services (e.g. cumulative compressed airflow demand for whole factory) and billing purposes (e.g. 15 minute interval of total electricity consumption as base of energy cost calculation) (criterion *realistic representation of consumption dynamics*). Following the previous criteria, interdependencies within the whole factory system need to be considered realistically. For example, energy and resource consumption as well as supply of supporting processes (e.g. technical building services) are directly depending on the state of the manufacturing system and vice versa. Just naming e.g. the total electricity consumption of compressors for a month is certainly a starting point. However, it does not support a detailed consideration of cause and effect in order to optimise the whole system or realistic allocation of consumption and costs (criterion *interdependencies with technical building services*).

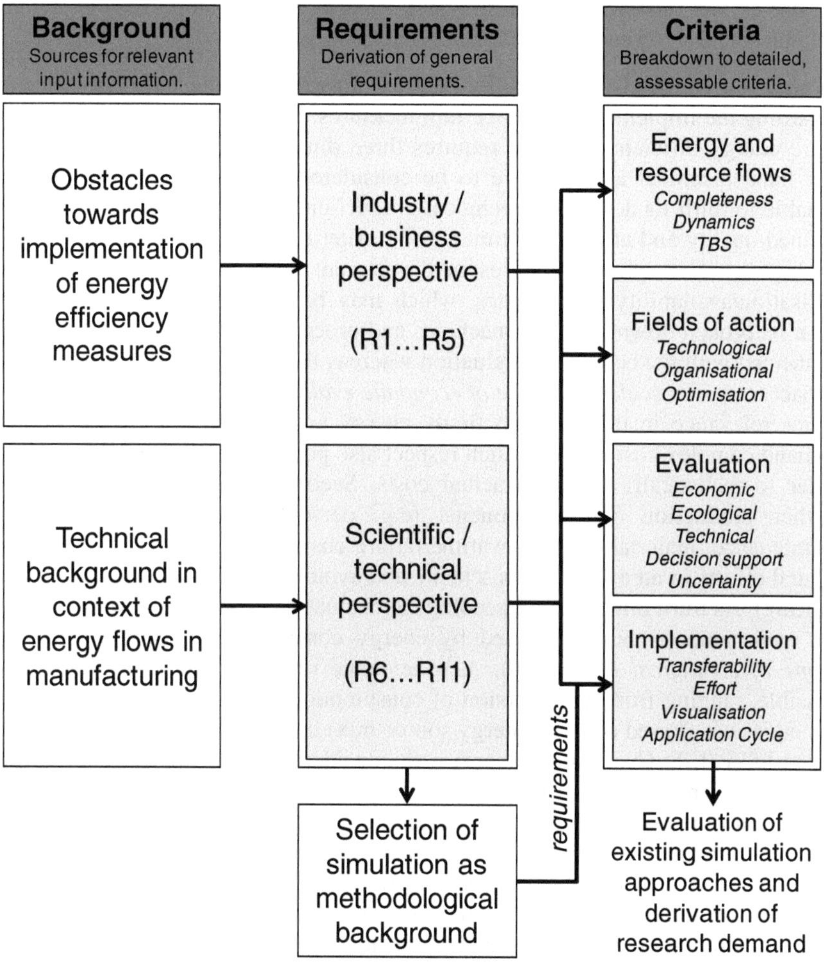

Fig. 36 Methodology for deriving requirements and criteria for the solution approach

Focused Fields of Action for Improvement: As pointed out above, diverse measures to improve energy efficiency in manufacturing systems are basically available. Simulation based approaches should be able to derive, analyse and/or evaluate all kinds of measures on different levels. This includes technological measures (criterion *technological measures*) like machine, TBS or manufacturing system design and control (e.g. impact of new machine concept in context of whole system, manufacturing system layout/logistics, installed load management to avoid energy peaks) as well as organisational measures (criterion *organisational measures*) like production planning (e.g. sequencing, order/lot sizes), maintenance strategies (e.g. preventive maintenance) or the management of employee behaviour (e.g. rules, operating procedures). Ideally, the usage of optimisation methods should be possible: going beyond conducting simulation experiments with pre-defined parameter sets this would allow to strive towards not just improved but actually optimal solutions (criterion *possibility of optimisation*).

Evaluation Methods: The results of an evaluation are the base for decisions regarding the implementation of certain measures. By definition, striving towards more energy efficient solutions requires three dimensions of evaluation. On the one hand technical aspects have to be considered to ensure that the system is capable to fulfil its designated technical production task (to deliver a product with defined quality and at the right time) at all (criterion *scale and scope of technical evaluation*). Example key figures in the present context are throughput times, utilisation/availability/quality rates, which may be available on system level or even be broken down to single machines and processes. These aspects are closely connected with the economic evaluation whereas this is a transformation into cost impact (criterion *scale and scope of economic evaluation*). Two main issues are of major relevance in this context: firstly energy and specifically electricity costs demand complex cost models which respect also peak surcharges and daily time in order to realistically calculate actual costs. Secondly, for the consideration of further production cost components (e.g. personnel costs for operation or maintenance, material costs, downtime/failure costs) it is desirable to put energy related efforts in an overall costs scheme and avoid negative side effects (e.g. cost shifting). As third dimension, an ecological evaluation is necessary, which reflects the environmental burden caused by energy consumption (criterion *scale and scope of ecological evaluation*). A wide range regarding the level of detail is possible, ranging from a conversion of consumption values to related greenhouse gas emissions (based on local energy source mix) up to detailed product or process based LCA (Life Cycle Assessment) with consideration of all input and outflows and diverse impact categories. As pointed out before in decision situations all evaluation dimensions need to be considered simultaneously, which may lead to conflicts of goals. To support decision appropriate methods to balance conflicts of goals and derive the most promising solutions from an integrated perspective is necessary (criterion *decision support*). As both are an advantage but also challenge applying simulation often involves the consideration of stochastic behaviour, e.g. when considering failure probabilities of machines, or volatile time distributions of operations or maintenance actions. This issue needs to be addressed to ensure the quality of results with either assuring that stochastic effects are not significant in the specific case or providing statistical confidence through sufficient simulation runs (criterion *consideration of uncertainty*).

Implementation: Besides rather content driven criteria some aspects regarding the embodiment of simulation based research approaches (referring to the challenges connected with simulations studies as described above) need to be addressed. Transferability is one example which stands for the broad applicability of the approach for different cases and purposes with reasonable effort (criterion *transferability*). This involves several perspectives: on the one hand not only one specific case shall be considered, e.g. the production of one selected product on a fixed production line. It should rather be possible to consider multi-product production with, ideally, flexible production process chains. Furthermore the solution should not be too company and industry specific but be easily applicable to other companies of the same or other industries. Somehow related to that issues like the possible adaption to other questions (e.g. using the same approach to improve manufacturing system and/or technical building services and/or product design while determining related environmental impact) or the necessary knowledge of users (easy involvement of other people possible?) will also be evaluated. Another crucial aspect is the effort in terms of time and costs for modelling and simulation (criterion *modelling and simulation effort*). Applying simulation methods demand specific and often expensive software tools as well as time to model, simulate (computing time) and evaluate the actual case. Being more visionary than realistic, an ideal approach should be based on free or reasonable priced standard software with fast simulation routines as well as appropriate methods to support easy and flexible design of the model (e.g. standard modules) and integrated/automated evaluation. Additionally, visualisation is of major importance while it enables to check the state of the simulation during runtime and supports the understanding of cause and effect within the system (criterion *visualisation*). It can be distinguished between the visualisation of the considered system itself displaying all things happening during runtime (e.g. material flow, machine states) and also of the relevant target criteria with proper key figures (and hints for improving the system, e.g. share of standby consumption) for evaluating the energy efficiency. To tab the full potential of simulation studies both perspectives should be displayed meaningfully and simultaneously during runtime. In the worst case, simulation runs are processed in the background without further visualisation and some results were generated which may even have to be transformed manually into relevant key figures. Certainly it should be mentioned that for some purposes (e.g. conducting various runs for parameter optimisation) a detailed visualisation of the production environment is not reasonable while unnecessarily leading to longer computing time. However it should be at least available as an option while strongly contributing to understand system behaviour. Finally, to support the correct and target oriented applicability, the simulation approach should ideally be embedded within an appropriate comprehensive methodology respectively application cycle (criterion *embedment within application cycle*). Starting from scratch without any specific prerequisite therewith, it shall be possible to get to promising measures to improve energy efficiency based on a verified/validated simulation model with appropriate scale and scope (e.g. definition of system boundaries and evaluation criteria, relevant energy and resource flows and possible data sources, level of detail and accuracy).

Table 4 gives an overview of all criteria with their specific range of value in a five step evaluation scheme. Each field (except if there is no fulfilment) counts for a quarter of a point and can sum up to a maximum of one (100% fulfilment) per criterion.

Table 4 Criteria for evaluation of research approaches

Criteria	○	+◔	+◔	+◔	+◔
Energy and Resource Flows					
Completeness	no flows	one flow	more than one flow	all internal	all external
Dynamics	single value	one flow state based	more than one state based	cumulative load profile for one flow available	cumulative load profiles for several flows available
TBS	not considered	consumption of one TBS subsystem considered	consumption of more than one TBS subsystem considered	interactions of one TBS subsystem considered	interactions of more than one TBS subsystem considered
Fields of action					
Technological	not considered	Machine improvement	TBS improvement	Prod. System improvement	integrated perspective
Organisational	not considered	PPC focus (depending on actual strength)		employee/behaviour focus (depending on actual strength)	
Optimisation	not considered	addressed and possible, not necessarily conducted		optimisation studies are conducted	
Evaluation					
Economic	not considered	conversion in costs	consideration of several energy carriers	realistic cost model including power peaks	complete cost model with different cost portions
Ecological	not considered	conversion to e.g. CO_2 one energy flow	conversion to e.g. CO_2 several flows	simplified LCA with selected flows/ impacts	full LCA based on LCI databases
Technical	not considered	output/production time considered		further aspects considered	
Decision Support	not considered	comparison/discussion		methods for integrated evaluation	
Uncertainty	not considered	addressed, not necessarily actually considered in detail		considered through appropriate methods	
Implementation					
Transferability	specific solution	branch spanning approach	approved for different manufacturing systems	appropriate to solve other questions	not necessarily expert usage
Effort	time and costs high	little expertise necessary	low software costs	low modelling time	low simulation time
Visualisation	not provided	material flow visualised in runtime	results in runtime	automatic processing of results at the end	meaningful processing for decision support
Application Cycle	not provided	comprehensive-ness ensured	validity ensured	support for systematic improvement	support for data acquisition

4.2 Evaluation of Relevant Research Approaches

Based on the previously described limitations 12 relevant research approaches were identified. They will be presented in the following. Beginning with some general information and a short description, each approach will be discussed based on the evaluation criteria described above. The evaluation can obviously just be based on information given in publicly available sources. This may involve that certain criteria cannot be assessed due to a lack of information. However, this will be indicated in the discussion.

Table 5 Evaluation of SIMTER approach developed by Heilala et al.

Main Author(s)	Juhani Heilala et al., Björn Johannsson et al. (SIMTER consortium)	
Institutions	VTT Technical Research Centre of Finland, Finland Product and Production Development, Chalmers University of Technology, Gothenburg, Sweden	
References	(Heilala et al., 2008; Lind et al., 2009)	
Software tools	3DCreate and 3DRealize of Visual Components, MS Excel	
Energy and Resource Flows		
Completeness	◑	all external energy sources considered, no internal flows
Dynamics	◑	state based energy consumption of machines, no cumulative load profiles
TBS	○	TBS not considered at all
Fields of action		
Technological	◔	mainly manufacturing system design is focused
Organisational	○	not considered
Optimisation	○	not addressed
Evaluation		
Economic	○	not considered
Ecological	◕	LCA based on LCI data planned; practically mainly energy related CO_2 emission considered

Table 5 (*continued*)

Technical	◔	mainly capacity considered in context of energy consumption evaluation
Decision Support	◑	Methods for an integrated evaluation of ergonomics, level of automation and environmental impact, but less in terms of energy efficiency
Uncertainty	○	not addressed
Implementation		
Transferability	◑	different branches and questions can be considered, unclear whether appropriate for flexible multi-product production, rather expert solution
Effort	◔	if modules in simulator available relatively fast modelling possible, expensive simulators and expertise necessary
Visualisation	◔	material flow visualised in 3D, due to model approach no energy related results during runtime, manual post-processing of data necessary, no support regarding meaningful visualisation of results
Application Cycle	○	not provided

The SIMTER project was carried out during 2007 and 2009 as a Finnish-Swedish cooperation (Lind et al., 2009). The project as a whole aimed at developing an integrated simulation tool simultaneously analysing the level of automation, the environmental impact and ergonomics of manufacturing systems as decision support for manufacturing system design engineers (Heilala et al., 2008). It focuses on the manufacturing system planning phase before first implementation and start-up. Thus issues like manufacturing system layout/bottleneck analysis are of major interest e.g. through varying the number and dimensioning of production equipment (e.g. cycle times, buffer sizes). Figure 37 shows the analysis flow of the environmental impact calculation sub-tool which is mainly relevant here.

Basically it is a combination of discrete event simulation and analytic calculation. The manufacturing system simulation is based on 3DCreate and 3DRealize of Visual Components and generates operation data (e.g. piece count, material used) and percentages of machine states (on, off, standby, under repair) as output. With this, after simulation a MS Excel based analytical calculation of environmental metrics (based on LCI data) is conducted, which includes impacts through energy and material consumption as well as emissions. This separation of simulation and evaluation procedure impedes the consideration of time dependent

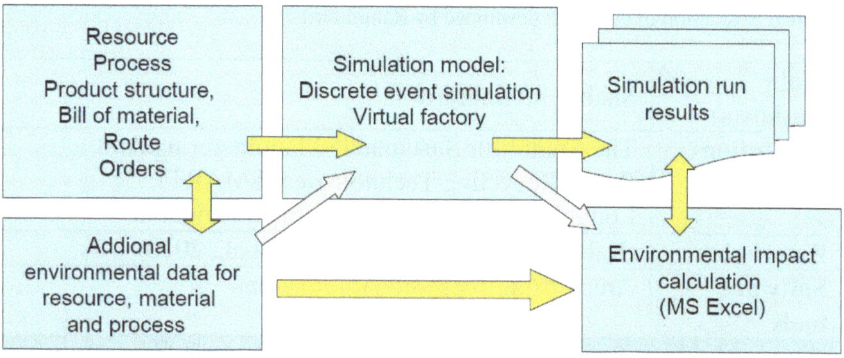

Fig. 37 Simplified analysis flow chart of SIMTER approach (Heilala et al., 2008)

cumulative load profiles and evaluating energy related cause and effect during runtime. The SIMTER approach does not require specific machine models (only an abstract simple state representation is necessary) and employs conventional discrete event manufacturing simulation tools so transferability as well as modelling effort in terms of time and cost are mainly restricted by properties, opportunities and limitations (e.g. library of standard machines, licence costs) of these tools. Case studies in different industries (automotive, food industry) underline the flexibility of the approach. However, each case considers only single product production situations with fixed production layout and no detailed production program (just one product continuously running through). The question arises whether this approach is also applicable for multi-product production with more complex and flexible production process chains with certain production planning and control.

Rahimifard presents a more product oriented approach for modelling embodied product energy (EPE, in kJ) during manufacturing. The consumption behaviour of machines and processes over time is not modelled; the energy is calculated per product based on equations in combination with metered data. Hence, a load profile of the whole factory system is not available. Rahimifard categorises energy consumption in manufacturing into two groups which sum up to the embodied energy of a product (Figure 38). Direct energy (DE) is required to manufacture the product in a specific process and can be further on, divided in theoretical energy (TE, physically necessary energy for actual value creation) and auxiliary energy (AE, supporting activities for the individual machine/process e.g. coolant pump). With available equations from production engineering TE can be calculated based on the characteristics of the product. AE needs to be determined through measurements. Additionally Rahimifard defines the Indirect Energy (IE) that is necessary to maintain the production environment (lighting, heating or ventilation) whereas different zones can be distinguished. The allocation of IE to the embodied product energy is based on the total IE consumption per hour of the specific zone

Table 6 Evaluation of approach developed by Rahimifard

Main Author(s)	Shahin Rahimifard	
Institutions	The Centre for Sustainable Manufacturing And Reuse/Recycling Technologies (SMART), Loughborough University, United Kingdom	
References	(Rahimifard, 2009; Rahimifard et al., 2010)	
Software tools	Arena from Rockwell Automations	
Energy and Resource Flows		
Completeness	◔	approach speaks of "energy" (with the unit kilo joule) in general, specific forms of energy are not mentioned or distinguished
Dynamics	○	consumption behaviour of machines is not modelled; energy is calculated per product based on equations in combination with metered data
TBS	◔	Energy consumption of TBS only considered in context of providing production conditions, certain interdependencies and/or the internal supply of different forms of energy are not addressed
Fields of action		
Technological	◔	fields of action are being revealed in general, feedback to product possible
Organisational	◔	PPC measures slightly addressed (e.g. batch sizes)
Optimisation	○	not considered
Evaluation		
Economic	○	no specific economic criteria considered
Ecological	○	no specific ecological criteria considered
Technical	◔	embodied product energy (EPE) in general as key criterion
Decision Support	◑	distinction between EPE portions as base for efficiency ratios and decisions
Uncertainty	○	not considered

Table 6 (*continued*)

Implementation		
Transferability	◔	basically flexible but specific equations and/or metered data necessary, application to more complex production scenarios questionable, rather expert approach
Effort	◔	Expensive simulation software as well as significant time for modelling (e.g. equations for TE) necessary if library not available
Visualisation	◑	material flows and relevant results basically available in runtime, within methodological restrictions manual generation of meaningful charts possible
Application Cycle	○	not provided

divided by the total number of products processed in one hour. Taking this into consideration the embodied energy over the whole process chains (several processes in different zones) for products can be calculated. Applicability directly depends on the availability of appropriate equations and metered data. To facilitate applicability while including the complexity of the problem, an energy simulation model was developed (based on simulation tool Arena from Rockwell Automations) which automatically calculates the energy values. However, dynamics of energy and resource consumption and also material flow are rarely considered in the presented cases; as also illustrated in (Rahimifard et al., 2010) the EPE could even be calculated statically. Certainly, there is potential when considering more complex cases and e.g. waiting times play a more important role. Whereas supporting decisions in operations management is possible, the simulation model shall explicitly help to derive improved design solutions in terms of "design for energy minimisation".

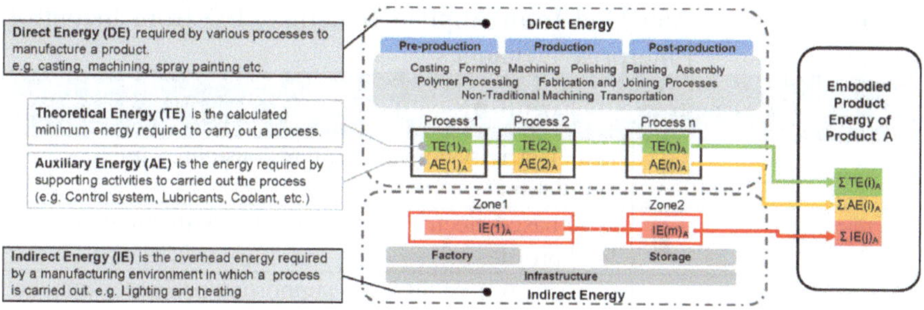

Fig. 38 The Embodied Product Energy framework for modelling energy flows during manufacture (Rahimifard et al., 2010)

Already in 2005 Solding et al. firstly presented their approach, which focuses on increasing the energy efficiency of Swedish foundries through applying energy aspects on discrete event simulation. The authors conducted several case studies of different foundries while individually extending conventional simulators like AutoMod or ED to depict the specific characteristics for each case (e.g. objective of case studies, scope of simulation study, manufacturing system structure, PPC aspects). In terms of energy carriers electricity is considered exclusively; state based consumption behaviour of machines is realised through additional programming within the base simulation tools. The additional overhead energy consumption of supporting processes is addressed (lighting, ventilation, space heating) and modelled as continuous consumer or depending on production machine working periods. Later publications introduced a tool for structured data gathering and transfer to the simulation environment. Improvement measures focus on production planning and control. Thereby, besides reducing the energy consumption, the consideration of energy peaks is of major interest while it may have significant cost impact – especially in energy intensive companies like metal foundries. The evaluation of peaks is possible as the simulation is able to generate cumulative electricity load curves (with appropriate time resolution) for the considered manufacturing system based on single machine consumption behaviour.

Table 7 Evaluation of approach developed by Solding et al.

Main Author(s)	Petter Solding, (Damir Petku, Patrik Thollander)	
Institutions	Department of Production, Energy and Environment Swedish Foundry Association, Jönköping, Sweden Production and Energy Technology, Swerea SWECAST AB, Jönköping, Sweden Vestas Guldsmedshyttan AB, Guldsmedshyttan, Sweden	
References	(Solding and Petku, 2005; Solding and Thollander, 2006; Solding et al., 2009)	
Software tools	AutoMod from Applied Materials, ED from Incontrol Enterprise Dynamics	
Energy and Resource Flows		
Completeness	◔	just electricity consumption is considered in simulation
Dynamics	◑	state based electricity consumption with cumulative load profile
TBS	◑	consumption of different supporting processes may be included but without detailed consideration of energy related interdependencies

Table 7 (*continued*)

Fields of action		
Technological	◔	manufacturing system improvement partly relevant in selected case studies
Organisational	◑	strong focus of coordination of processes through PPC
Optimisation	○	not really addressed, mentioned in outlook of most recent paper
Evaluation		
Economic	◑	diverse issues of electricity pricing are addressed and seemingly included (not in detail shown in case studies, maybe due to confidentiality issues)
Ecological	○	no ecological evaluation
Technical	◔	productivity (number of parts) as indicator for technical performance
Decision Support	◔	technical and energy related issues are discussed, no actual decision support
Uncertainty	○	not addressed
Implementation		
Transferability	◔	quite specific solution for each case ("individually built"), high expertise necessary
Effort	○	Expensive simulation software as well as significant time and expertise for modelling (specifics of each case need to be addressed) necessary
Visualisation	◔	material flow (apparently) visualised in runtime, detailed results (including cumulative load profile) just available after simulation run, no processing for systematic decision support observable
Application Cycle	◑	software aided data collection was introduced which also supports to ensure the comprehensiveness of the energy study

Weinert et al. developed a concept for "energy-aware production planning and control" (Chiotellis et al., 2009) based on so called energy blocks. They argue that the specific load profiles of production machines can be separated into distinctive classes ("energy blocks") based on operational states (Figure 39).

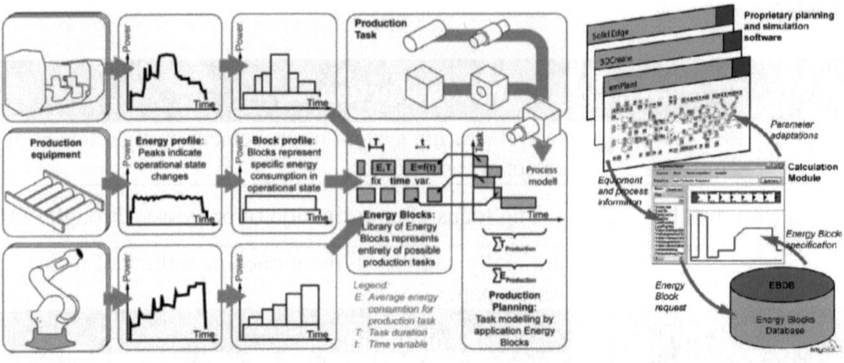

Fig. 39 Planning methodology based on energy blocks and related interface to simulation software (Chiotellis et al., 2009)

The energy consumption of each block (exclusively focusing on electricity) can be described through mathematical equations. Therewith energy consumption data is made processable and can be used to model specific production tasks (process chains) through application and combination of the relevant energy blocks. Measured and processed energy blocks are stored in a specific database. In a further step this data was combined with discrete event simulation. As shown in Figure 39 conventional simulation tools like 3DCreate or emPlant (Plant simulation) can be used to simulate manufacturing systems and deliver relevant data like operation states or process times. Within a calculation module this data is merged with involved energy blocks – hence, the calculation of the energy consumption is enabled.

In 2007, in his dissertation Junge presented a unique approach for considering energy consumption in manufacturing systems, strongly focusing on the specific case of injection moulding. The considered manufacturing system is modelled with Simflex 3D, a material flow simulation software that is an own development of the involved research institution. State based electricity consumption is directly handled within the simulation environment. However, in injection moulding additional, TBS related, energy consumption for air processing (possible emissions of harmful substances) and heating is very relevant. To tackle this issue Junge uses TRNSYS, which is an established tool for simulation air/heat flows in buildings normally applied by civil engineers. Junge couples both material flow and building simulation through defined interfaces (Figure 40): time respectively state depending heat and particle emissions of injection moulding machines simulated in Simflex are being transferred to TRNSYS where the additional demand of electricity of oil for heating and air conditioning is calculated dynamically (also including influencing variables like the local climate).

Table 8 Evaluation of approach developed by Weinert et al.

Main Author(s)	Nils Weinert, Stylianos Chiotellis et al.	
Institutions	Department of Machine Tools and Factory Management (IWF), Chair of Assembly Technology and Factory Management, Technische Universität Berlin, Germany	
References	(Weinert et al., 2009; Weinert, 2010; Chiotellis et al., 2009; Fleschutz et al., 2010)	
Software tools	emPlant/3dCreate	
Energy and Resource Flows		
Completeness	◔	just electricity is considered as of now
Dynamics	◔	state based modelling of electricity consumption, however no cumulative load profiles
TBS	○	focus on process chain, no consideration of any supporting processes
Fields of action		
Technological	◔	manufacturing system design is mainly addressed
Organisational	◔	PPC issues partly addressed, but not main focus
Optimisation	○	not considered
Evaluation		
Economic	◑	cost calculation (including peak prices) is addressed
Ecological	○	not considered in original approach by Weinert et al.
Technical	◔	throughput time is considered
Decision Support	○	not considered in original approach by Weinert et al.
Uncertainty	○	not addressed

Table 8 (*continued*)

Implementation		
Transferability	◑	basically branch spanning just depending one data availability, different studies possible, application to complex production scenarios unclear
Effort	◔	if modules in simulator available relatively fast modelling possible, expensive simulators and expertise necessary
Visualisation	◔	material flow shown in runtime but no energy related results, manual processing of results necessary
Application Cycle	○	not provided

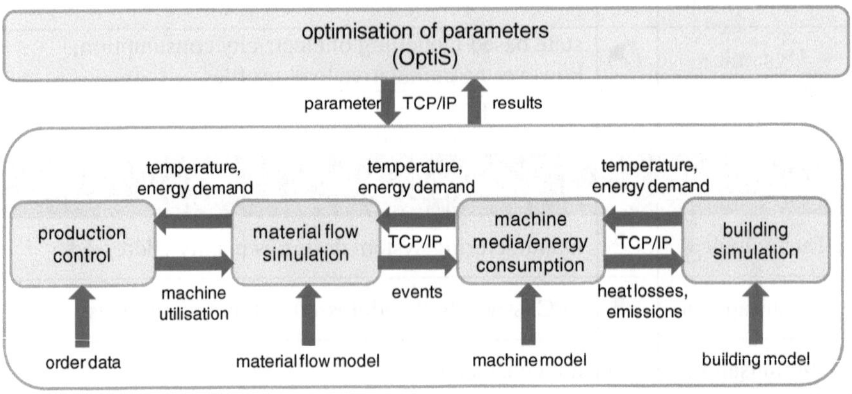

Fig. 40 Conceptual framework of simulation approach based on (Junge, 2007)

Diverse scenarios for different production strategies but also for different locations (with different climate) are considered. The derivation of promising strategies is supported by a tool for support parameter optimisation (OptiS, own development). For evaluation, Junge introduces a multi-criteria approach that includes diverse logistical objectives as well as an extensive cost model going beyond just energy cost calculation.

Hesselbach et al. introduced a more complex simulation approach which explicitly considers the interdependencies between production factors and technical building services. It was developed within the research project ENOPA ("Energy efficiency through optimised coordination of production and TBS"). It aims at an optimised dimensioning and efficient process control of technical

Table 9 Evaluation of approach developed by Junge

Main Author(s)	Mark Junge	
Institutions	Fachgebiet Umweltgerechte Produkte and Prozesse (upp), Universität Kassel, Germany	
References	(Junge, 2007; Hesselbach and Junge, 2005)	
Software tools	Simflex 3D, TRNSYS, Optis	
Energy and Resource Flows		
Completeness	◕	almost all relevant energy flows considered, except compressed air
Dynamics	●	all flows sufficiently dynamically considered, cumulative profile available
TBS	◑	heating/AC is considered in interaction with manufacturing system
Fields of action		
Technological	○	not in focus, no technological changes on machines, manufacturing system or TBS considered
Organisational	◑	very strong focus on PPC
Optimisation	●	parameter optimisation studies were conducted
Evaluation		
Economic	●	holistic cost calculation including several cost components
Ecological	○	not considered
Technical	●	diverse logistic/technical key figures (e.g. utilisation, throughput time, stocks) are considered
Decision Support	◑	Comparison/discussion but no actual decision support
Uncertainty	●	is addressed through proving the independence from stochastic effects

Table 9 (*continued*)

Implementation		
Transferability	○	very specific solution which demands expert knowledge, considering a new manufacturing system would need a completely new modelling
Effort	○	two simulation tools involved, extensive time needed for modelling and simulation
Visualisation	◑	material flow simulation in 3D during runtime, manual processing for meaningful visualisation necessary
Application Cycle	○	not provided

building services with respect to dynamic influences from production and, vice versa, an energy efficient production management which includes restrictions and opportunities given by technical building services. A holistic simulation approach is used, which integrates and dynamically couples different simulation tools for all relevant layers of the problem through certain interfaces (Figure 41):

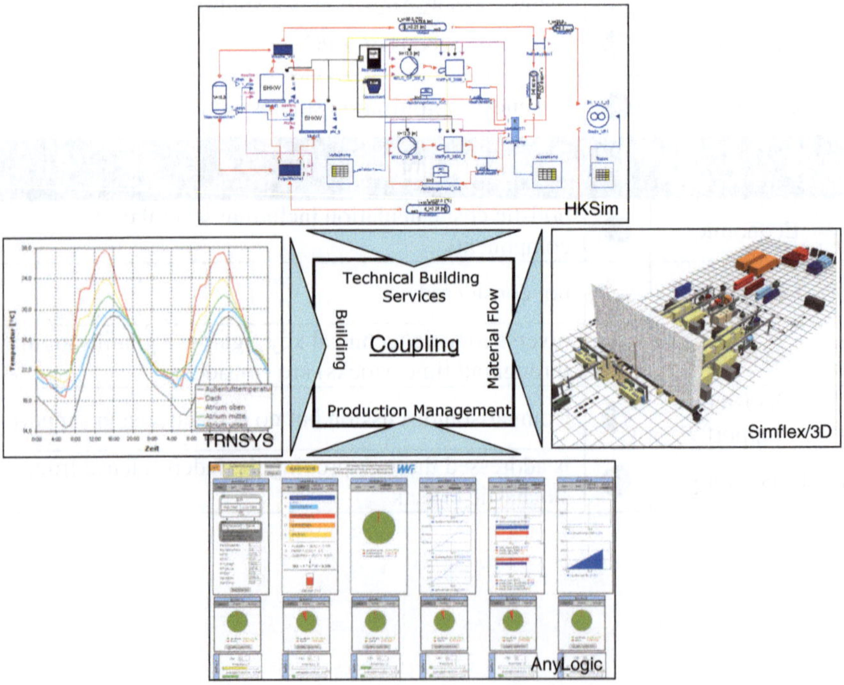

Fig. 41 Conceptual framework of ENOPA coupled simulation approach (Hesselbach et al., 2008b)

- Technical Building Services: the simulation tool HKSim was invented by the company Imtech Deutschland GmbH & Co KG. It allows simulating dynamically the generation, circuitry and consumption of energy and media within technical building services. Thereby HKSim supports the design and control of these complex systems (Wischhusen et al., 2003).
- Building (climate): the (plant) building itself is the place where technical building services (e.g. air conditioning) and production (e.g. waste heat, exhaust air) interact, which is specifically important when a defined production environment is crucial. To simulate the thermal processes TRNSYS as an established tool in this field is used.
- Production Machines / Material Flow: the material flow within the manufacturing system can be simulated with SIMFLEX/3D, which was developed by the University of Kassel
- Production Management: the software AnyLogic from XJTek as multi-paradigm simulation platform is used to simulate the influence and interdependencies of production management measures (e.g. production program, production control, production strategies) within the process chain from a more abstracted perspective. Promising scenarios are being transferred to the more detailed coupled simulation.

The simulation tools themselves are not new; they are all established solutions within their specific fields of application. The most important aspect is the shift away from an isolated usage to an integrated approach by coupling them. This allows a realistic and holistic modelling of the whole system of production and technical building services. Under consideration of all relevant interdependencies and dynamics of the system, different scenarios can be analysed and evaluated with both economic and ecological criteria. As a result, this enables the derivation of optimal solutions from a global perspective.

Table 10 Evaluation of EnoPA approach developed by Hesselbach et al.

Main Author(s)	Jens Hesselbach, Lars Martin, Christoph Herrmann, Sebastian Thiede, Rüdiger Detzer, Bruno Lüdemann (EnoPA consortium)
Institutions	Fachgebiet Umweltgerechte Produkte und Prozesse, Universität Kassel, Germany Product- and Life-Cycle-Management Research Group, Institute of Machine Tools and Production Technology (IWF), Technische Universität Braunschweig Imtech Deutschland GmbH
References	(Hesselbach et al., 2008a; Hesselbach et al., 2008b)
Software tools	Simflex 3D, TRNSYS, HKSim, (AnyLogic)

Table 10 (*continued*)

Energy and Resource Flows		
Completeness	◕	almost all relevant energy flows considered, except compressed air
Dynamics	●	all flows sufficiently dynamically considered, cumulative profile available
TBS	●	heating, AC, cooling media as well as steam generation is considered in interaction with manufacturing system
Fields of action		
Technological	◕	manufacturing system and TBS consideration with integrated perspective
Organisational	◑	effects of PPC measures are considered
Optimisation	○	not focused
Evaluation		
Economic	◕	realistic cost calculation for different energy carriers
Ecological	◑	conversion to CO_2 with distinction of different countries
Technical	◑	production time/output as main technical variables
Decision Support	◕	comparison of different goals, discussion of integrative evaluation
Uncertainty	○	not addressed
Implementation		
Transferability	○	very specific solution which demands expert knowledge, considering a new case would need a completely new modelling
Effort	○	at least three simulation tools involved, extensive time needed for modelling and simulation
Visualisation	◑	material flow simulation in 3D during runtime, manual processing for meaningful visualisation necessary
Application Cycle	○	not provided

In 2009, Fraunhofer IPA introduced their Total Energy Efficiency Management (TEEM) concept, which aims to provide support for energy data metering, visualisation, monitoring, and systematic derivation of improvements. As one aspect, discrete event simulation is used to evaluate and derive measures towards energy efficiency. Available literature focus on one specific example to describe procedure and potentials of TEEM, namely the process chain injection moulding, painting, drying and assembly for plastics. This process chain was also modelled in Plant Simulation from Siemens UGS with consideration of main input energy carriers, electricity, compressed air and process heat. While being relevant in context of energy consumption for painting and drying also air ventilation was considered with specific models, which enable to evaluate improvements in this subsystem. TEEM simulation provides an economic evaluation of improvement measures including both energy costs (simplified calculation just based on consumption) and necessary investment in order to calculate amortisation time.

Table 11 Evaluation of approach developed by Fraunhofer IPA

Main Author(s)	e.g. Markus Hornberger (TEEM-Total Energy Efficiency Management)	
Institutions	Fraunhofer IPA, Stuttgart, Germany	
References	(Hornberger, 2009, Hornberger, o. J.; Wahren, o. J.)	
Software tools	Plant Simulation 8.1	
Energy and Resource Flows		
Completeness	◕	relevant energy flows are considered but no details regarding origin of compressed air and process heat from energy perspective
Dynamics	●	single machine as well as cumulative load profiles for all considered energy flows are obviously available
TBS	◑	consideration of air ventilation for selected processes, no further details regarding generation of compressed air and process heat
Fields of action		
Technological	◑	focus on machine and TBS improvement, manufacturing system rather fixed
Organisational	◔	PPC is addressed but not in main focus
Optimisation	○	not addressed

Table 11 (*continued*)

Evaluation		
Economic	◑	simplified conversion to energy costs with same price rate for all energy carriers, investment/amortisation of measures included
Ecological	○	not considered
Technical	◑	production time/rate considered
Decision Support	◑	comparison of different target variables and calculation of amortisation for but e.g. not consideration of technical issues in this case
Uncertainty	○	not addressed
Implementation		
Transferability	◔	basically yes, but not proved so far and always quite specific solution
Effort	○	Expensive licence, time and expertise for specific model necessary
Visualisation	◑	Very little information available but seemingly material flow and possible result visualisation in runtime, manual post-processing for meaningful charts necessary
Application Cycle	◑	Embedded in TEEM method which shall support in terms of data acquisition and systematic derivation of improvements

Löfgren published first work on a LCA oriented discrete simulation approach which concentrates on the complex manufacturing system of one specific Swedish company (SKF). The idea is to dynamically generate profiles in terms of energy consumption but also material losses by means of simulation in order to conduct LCA studies as base for improving the system. Besides manufacturing system improvement, an important motivation is to show how the environmental outcome can be influenced by employee behaviour.

Johansson et al. were also members of the aforementioned SIMTER consortium. While having extended the SIMTER approach in some important issues the evolved approach will be considered here separately as well. Johansson et al. focus on the ecological evaluation of manufacturing systems. The functional principle – modelling with conventional simulation software and manual processing in MS Excel afterwards – is quite similar to SIMTER whereas AutoMod was used here. In further publications they presented different case

Table 12 Evaluation of approach developed by Löfgren

Main Author(s)	Birger Löfgren	
Institutions	Chalmers University of Technology, Sweden SKF Sverige AB	
References	(Löfgren, 2009)	
Software tools	?	
Energy and Resource Flows		
Completeness	◑	several important flows are considered, not necessarily comprehensive
Dynamics	◑	state based modelling for all flows, not clear whether cumulative profiles are available
TBS	◑	consumption of compressed air and fluids are considered but simply converted to connected electricity consumption, no interactions
Fields of action		
Technological	◑	machine and manufacturing system perspective
Organisational	◔	PPC not in focus, but influence of employee behaviour is addressed
Optimisation	○	not considered
Evaluation		
Economic	○	no economic evaluation
Ecological	◕	complete LCA planned, however, just consideration of energy as of now
Technical	◔	Output is mainly considered
Decision Support	◑	comparison of a calculation of efficiency ratios to balance
Uncertainty	◔	necessity was addressed but no studies conducted

Table 12 (*continued*)

Implementation		
Transferability	◔	basically as idea certainly transferable, however very specific solution
Effort	◔	while using relatively easy process models, there is less effort for modelling
Visualisation	-	no information available
Application Cycle	○	not considered

Table 13 Evaluation of approach developed by Johansson et al.

Main Author(s)	Björn Johansson et al.	
Institutions	Product and Production Development, Chalmers University of Technology, Gothenburg, Sweden	
References	(Johansson et al., 2009b; Johansson et al., 2009a) (juice)	
Software tools	AutoMod, MS Excel	
Energy and Resource Flows		
Completeness	◑	all external energy sources considered, no internal flows
Dynamics	◑	state based energy consumption of machines, no cumulative load profiles
TBS	○	TBS not considered at all
Fields of action		
Technological	◔	mainly manufacturing system design is focused
Organisational	◔	order sizes and sequencing as important means in case study
Optimisation	◕	optimisation study was conducted

Table 13 (*continued*)

Evaluation		
Economic	○	not considered
Ecological	◕	LCA based on LCI data planned; practically mainly energy related CO_2 emission considered
Technical	◑	throughput times as well as output is considered
Decision Support	◑	comparison of target variables, efficiency ratios used
Uncertainty	◔	addressed in case study, but not explained in detail
Implementation		
Transferability	◑	different branches and questions can be considered, unclear whether appropriate for flexible multi-product production, rather expert solution
Effort	◔	if modules in simulator available, relatively fast modelling possible, expensive simulators and expertise necessary
Visualisation	◔	material flow visualised in 3D, due to model approach no energy related results during runtime, manual post-processing of data necessary, no support regarding meaningful visualisation of results
Application Cycle	○	not provided

studies (food/automotive) and underlined the flexibility of the approach. In comparison to the original work, the authors introduced PPC measures and optimisation within one study and also slightly addressed the issue of uncertainty with the possible range of results.

Dietmair and Verl consider the issue from a machine control oriented perspective and present a Petri-Net based approach to predict the energy (electricity) consumption of machine tools. In several publications they successfully proved that based on detailed state models, individual machine energy consumption (with supply units like coolant pumps) can be described with high accuracy. This enables to optimise process respectively machine control, identify critical components or realistically evaluate different machine alternatives. In a further step they argue that the same principle can basically also be applied to model manufacturing system with several connected machines (Figure 42) in order to support tactical and strategic manufacturing system planning.

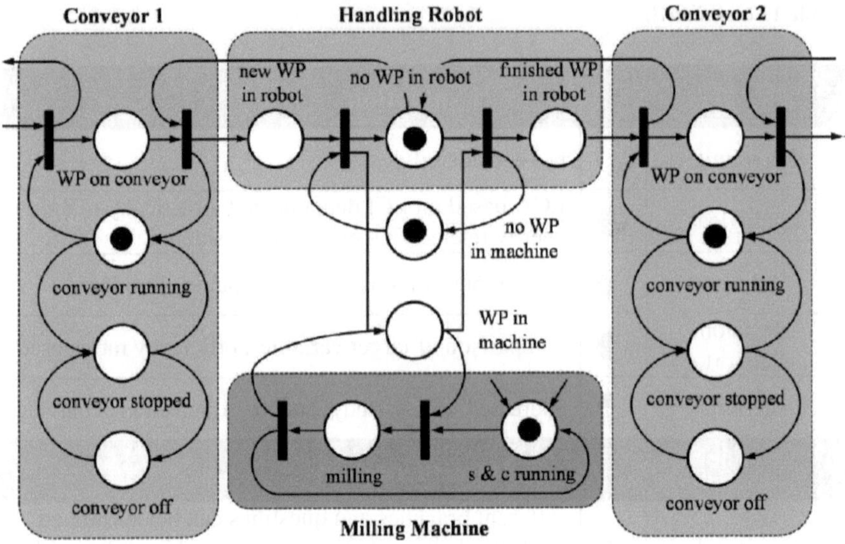

Fig. 42 High accuracy modelling of aggregate systems referring to (Dietmair and Verl, 2009)

Table 14 Evaluation of approach developed by Dietmair and Verl

Main Author(s)	Anton Dietmair, Alexander Verl	
Institutions	Institute for Control Engineering of Machine Tools and Manufacturing Units (ISW), Universität Stuttgart, Stuttgart, Germany	
References	(Dietmair and Verl, 2009; Dietmair and Verl, 2010)	
Software tools	?	
Energy and Resource Flows		
Completeness	◔	as of now just electricity consumption is considered
Dynamics	◑	very accurate state based modelling of electricity consumption
TBS	○	no consideration of TBS

Table 14 (*continued*)

Fields of action		
Technological	◑	strong focus on machine improvement, manufacturing system improvement addressed but some important questions remain unsolved as of now (e.g. material flow handling, production planning and control implementation)
Organisational	◔	organisational aspects like scheduling shortly addressed
Optimisation	◔	optimisation possibility on machine level addressed
Evaluation		
Economic	○	no economic evaluation
Ecological	○	no ecological evaluation
Technical	◔	Issues like production time slightly addressed
Decision Support	○	no discussion of different target objectives
Uncertainty	○	not addressed
Implementation		
Transferability	◑	basically flexible and scalable approach which can be used to tackle diverse questions, not clear whether applicable to different manufacturing systems
Effort	◔	Detailed modelling demands time and expertise, relatively fast simulation time assumed due to reduced approach
Visualisation	◔	No material flow visualisation in runtime, very detailed diagrams for analysis can be derived
Application Cycle	○	not provided

Material flow networks are a “special form of Petri-Nets and can be used to model material and energy flows in multi-stage manufacturing systems, including its environmental aspects” (Wohlgemuth et al., 2006) and are an important base for environmental management information systems (EMIS). Thereby transitions (material or energy transformations calculated based on equations/sub-models, shown as circle in diagram notation), places (inventories, squares) and flows (arrows) are being distinguished (see Figure 43). Material flow networks “serve as accounting systems for material and energy flows” (Wohlgemuth et al., 2006) and involve a set of rules like the principle of double-entry-bookkeeping of flows. However, cumulative values for defined intervals are typically used in transitions. Material flow networks do not focus on describing the time dependent behaviour and interactions of production processes and process chains in detail. Against this background Wohlgemuth et al. aim at combining material flow networks with discrete event simulation. Two different concepts were pursued so far: on the one hand the transformation processes within one transition can be calculated based on an inventory simulation which is being executed when the transition occurs. The results of the simulation are directly transferred to the material flow network software (e.g. Umberto by IFU Hamburg GmbH) and serves as output of the considered transition (see Figure 43). While being able to depict dynamics within one transition this concept does not allow the inclusion of interactions between different transitions of the system. As next evolutionary step Wohlgemuth et al. developed an approach, which combines discrete event material flow simulators like em-plant or Milan with material flow network software like Umberto. The production process chain is completely modelled within the simulation environment and experiments can be conducted. Per command the whole simulation model structure, as well as the results of the conducted experiment, are automatically transferred into Umberto. Based on the specific sub-models of the simulation all transitions, places, flows and result values can directly be used in order to e.g. conduct detailed economic or environmental evaluation within the material flow network software environment. A separate modelling of simulation model and material flow network is not necessary.

In 2007, the German based company Siemens AG overtook Tecnomatics Technologies Ltd./UGS Corporation as the developers of eM-Plant (formerly Simple++), which is a well-known and quite popular material flow simulator. In 2005 it was renamed to Tecnomatix Plant Simulation already. Plant Simulation can be clearly classified to discrete event simulation and is a typical material flow simulation software which supports the design and control manufacturing systems while e.g. revealing bottlenecks through simulation. While strongly focusing on the material flow itself supporting processes are typically not considered in detail. Continuous flows (like energy flows in interaction with technical building services) can hardly be depicted. Energy consumption of technical equipment is not an issue as of now. Being one of the standard tools for material flow simulation, some of the presented approaches use Plant Simulation as base for more detailed considerations of energy consumption. However, Siemens themselves strive to integrate energy consumption in more detail directly into their tool. Unfortunately very scarce information is available publicly as of now. On the

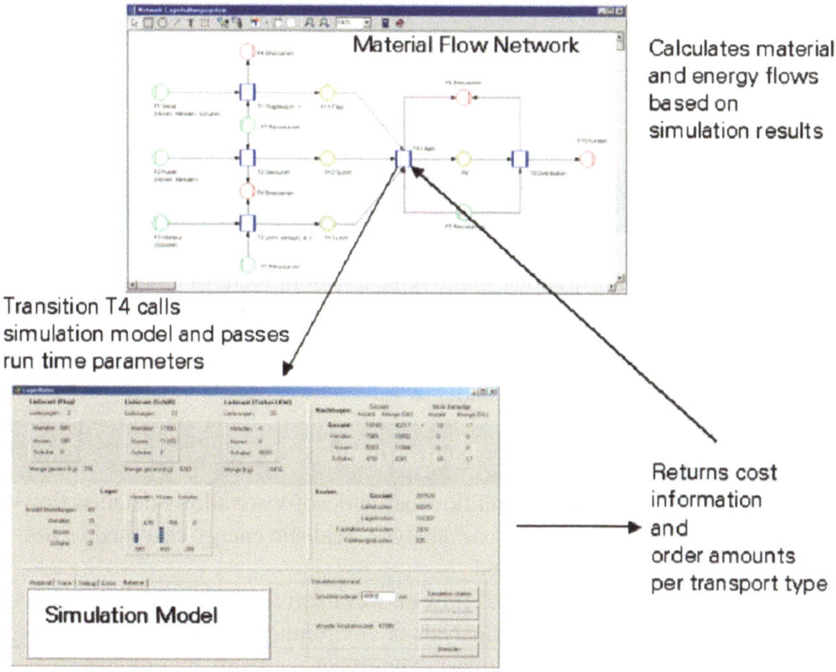

Fig. 43 Linking a Discrete Event Inventory Simulation to a Material Network (Wohlgemuth et al., 2006)

Table 15 Evaluation of approach developed by Wohlgemuth et al.

Main Author(s)	Volker Wohlgemuth	
Institutions	FHTW Berlin, Department of Technics II, Industrial Environmental Informatics, Berlin, Germany	
References	(Wohlgemuth, 2005; Wohlgemuth et al., 2006)	
Software tools	MILAN/emPlant, Umberto	
Energy and Resource Flows		
Completeness	●	basically very comprehensive view on energy and resource flows possible through energy and material flow network logic
Dynamics	○	due to the inner logic all consumption pattern are modelled per piece and not state-based

Table 15 (*continued*)

TBS	◑	not clearly shown but consumption of supporting processes can probably be considered, time-based interactions not
	Fields of action	
Technological	○	while rather approach for accounting and evaluation actual improvements not in focus of consideration
Organisational	○	while rather approach for accounting and evaluation actual improvements not in focus of consideration
Optimisation	○	not considered
	Evaluation	
Economic	◕	material flow network software allows detailed flow-based cost analysis, realistic energy cost models not included
Ecological	●	material flow network software allows detailed flow-based analysis of environmental impact
Technical	◔	not in focus of approach, certainly by-product of simulation studies
Decision Support	○	not explicitly offered
Uncertainty	○	not considered, in material flow software just one sample can be considered
	Implementation	
Transferability	○	very specific simulation sub-models and expert knowledge are necessary for considered case
Effort	○	high efforts and expertise for specific modelling, two tools necessary
Visualisation	◕	material flow shown in simulator, meaningful visualisation of results at the end of simulation
Application Cycle	○	not provided

Table 16 Evaluation of approach developed by Siemens

Main Author(s)	Siemens PLM Software/UGS Corporation/Tecnomatics Technologies	
Institutions	Siemens AG	
References	(Siemens AG, 2010)	
Software tools	Plant Simulation	
Energy and Resource Flows		
Completeness	◔	just electricity is considered
Dynamics	◔	state based consideration of electricity consumption (idle, load, start-up)
TBS	○	not considered
Fields of action		
Technological	◔	improvement of manufacturing system considered
Organisational	○	not considered
Optimisation	○	certainly possible but not considered as of now
Evaluation		
Economic	○	no economic evaluation
Ecological	○	no ecological evaluation
Technical	◑	throughput as main criterion
Decision Support	◑	short discussion of throuphput in relation to energy consumption
Uncertainty	○	not considered

Table 16 (*continued*)

Implementation		
Transferability	◑	basically transferable to all Plant Simulation application cases with certain limitations (e.g. flexible manufacturing systems)
Effort	◑	no second software and model necessary, "one-stop-solution"
Visualisation	◑	material flow in runtime, automatic processing of results
Application Cycle	○	not provided

Siemens PLM webpage only a very simple case of a conveyor is shown where production performance and connected energy consumption are considered simultaneously. However it illustrates the technical feasibility of the direct integration as well as the possible direction of development.

4.3 Discussion and Comparison

In the following chapter identified research approaches will be compared and research focuses as well as possible gaps are being discussed. This allows the later derivation of necessary research work in the field.

As base for further discussion Table 17 shows the summarised evaluation of all approaches in terms of the criteria. Although all approaches aim at a similar general objective – the simulation of the energy consumption of manufacturing system – their individual specifications are evidently very different.

In the next step the evaluation was transformed into quantitative values (each quarter counts for a value of 0.25 summing up to a maximum of 1.0), which allows a more detailed comparison of the fulfilment of criteria. Based on this transformation Figure 44 shows the average value (overall 12 considered approaches) for each criterion as well as overall criteria (dashed vertical line). Additionally the range is depicted as average of the six highest and lowest values in each category. Table 17 also contains the average value of all columns (approach based analysis). Based on this analysis several findings can be stated:

- Having in mind that 4.0 is the maximum and ideal value of each criterion it becomes clear that **there is generally significant room for improvement in all areas** towards the vision of a comprehensive integration of energy and resource flows into simulation based planning procedures.

- Some approaches fulfil certain criteria very well – however they involve significant drawbacks in other areas. There is **no approach with balanced and high fulfilment of all criteria**.
- Criteria completeness (of energy and resource flows) and dynamics are fulfilled higher than the average (still at relatively low level though). Technical variables are usually considered for evaluation, often in combination with other economic and/or ecological variables. In addition the criterion visualisation is fulfilled above average.
- On average some specifically critical issues can be identified: the lacking consideration of uncertainty and optimisation studies, the lack of an application cycle which supports implementation as well as the significant effort that is necessary to setup an energy oriented simulation of manufacturing systems.

Table 17 Comparison of evaluation results

	Heilala et al.	Rahimifard	Solding et al.	Weinert et al.	Junge	EnoPA	TEEM	Löfgren	Johannsson	Dietmair	Wohlgemuth	Plant Simulation
Energy and Resource Flows												
Completeness	◑	◔	◔	◔	◕	◕	◕	◑	◑	◔	●	◔
Dynamics	◑	○	◑	◔	●	●	●	◑	◑	◔	○	◔
TBS	○	◔	◑	○	◑	●	◑	◑	○	○	◑	○
Fields of action												
Technological	◔	◔	◔	◔	○	◕	◑	◑	◔	◔	○	◔
Organisational	○	◔	◑	◔	◑	◑	◔	◔	◔	◔	○	○
Optimisation	○	○	○	○	●	○	○	○	◕	◔	○	○
Evaluation												
Economic	○	○	◑	◑	●	◕	◑	○	○	○	◕	○
Ecological	◕	○	○	○	○	◑	○	◕	◕	○	●	○
Technical	◔	◔	◔	◔	●	◑	◑	◔	◑	◔	◔	◑
Decision Support	◑	◑	◔	○	◑	◕	◑	◑	◑	○	○	◑
Uncertainty	○	○	○	○	●	○	○	◔	◔	○	○	○
Implementation												
Transferability	◑	◔	◔	◑	○	○	◔	◔	◑	◑	○	◑
Effort	◔	◔	○	◔	○	○	○	◔	◔	◔	○	◑
Visualisation	◔	◑	◔	◔	◑	◑	◑	-	◔	◔	◕	◑
Application Cycle	○	○	◑	○	○	○	◑	○	○	○	○	○
Average	0.25	0.18	0.27	0.18	0.52	0.47	0.38	0.32	0.35	0.20	0.28	0.22

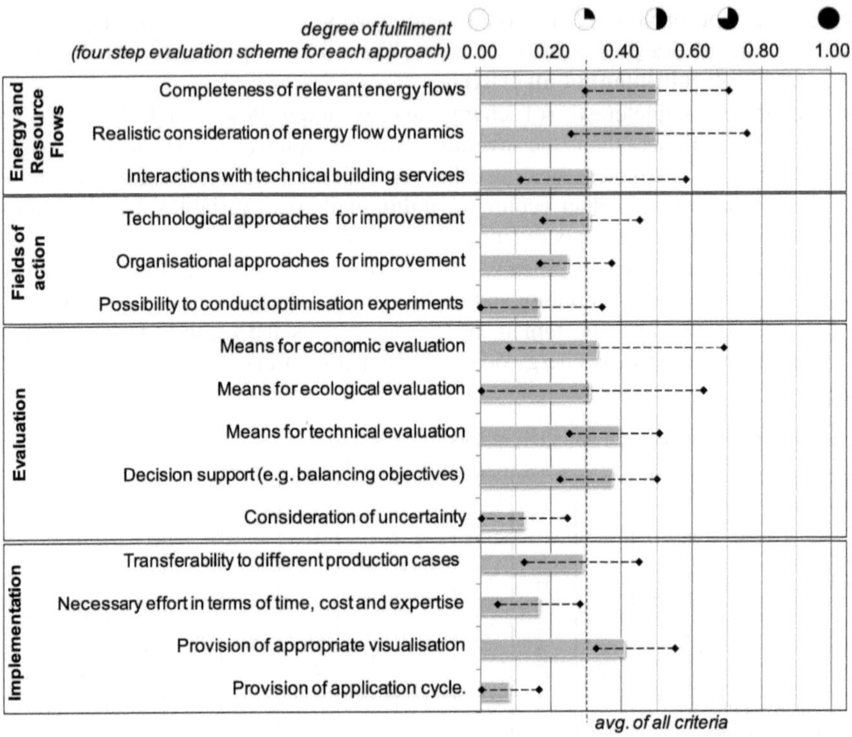

Fig. 44 State of research - degree of fulfilment regarding identified criteria towards energy oriented simulation

Based on the detailed analysis of the research approaches three different general paradigms for simulating energy flows in manufacturing systems can be identified (Figure 45):

- In paradigm A the manufacturing system is simulated through (typically commercial) a discrete event based material flow simulator. While these tools originally neglect necessary possibilities to consider energy flows, relevant information (e.g. operation states of machines) is transferred to a static evaluation application through a defined interface after the simulation run is finished. Through combination with energy related information (e.g. consumption in different states) an evaluation with respect to energy consumption can be conducted.
- Paradigm B depicts the coupling of different simulation approaches (plus optional evaluation) in order to respect the dynamic nature of energy flows as well as the complex interactions taking place in different sub-systems of a factory. Examples are the combination of material flow simulation with building (heat flow), simulation (Junge) or the EnoPA approach which additionally integrates technical building services simulation and a separate evaluation module.

- Paradigm C describes the direct extension/integration of material flow simulation with energy related flow information and evaluation schemes within one simulation environment (e.g. Siemens Plant Simulation).

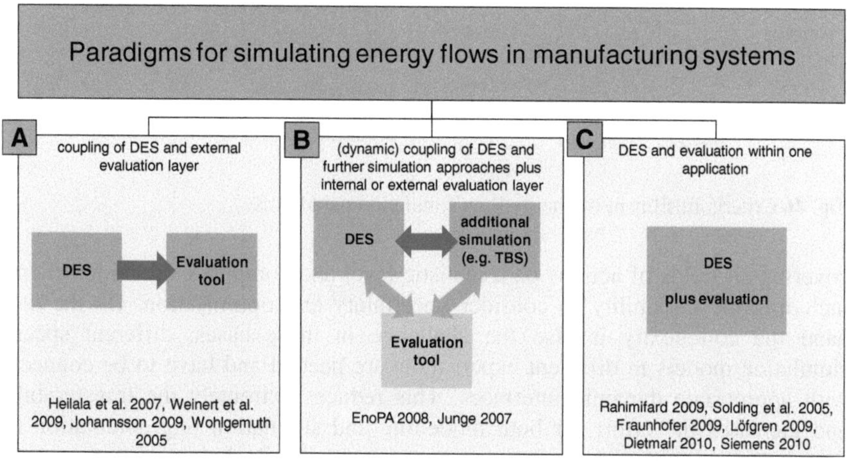

Fig. 45 Identified paradigms for simulating energy flows in manufacturing systems based on discrete event simulation (DES)

In the next step the analysis of all research approaches allows to derive characteristics of each of those paradigms. Figure 46 shows the results of this consideration. Due to its specific inherent logic each paradigm shows characteristic profiles in terms of the considered criteria set.

On average, simulation approaches classified as paradigm A (interface between material flow simulation and ex-post evaluation) offer relatively good coverage of manifold energy and resource flows, the possibility of extensive and distinctive evaluation systems as well as the low additional effort for modelling/simulation and good transferability. The main reason is the strict separation of simulation and evaluation which allows an uncomplicated and extensive design of the evaluation environment, which is also independent from specific processes and simulation models. As a disadvantage, through this isolated consideration certain energy oriented dynamics and interdependencies cannot or hardly be depicted (e.g. often no cumulative load profiles on system layer, missing energy flow induced events and interactions). This also reduces the degree of freedom in terms of considering fields of action towards better energy and resource efficiency. Additionally, simulation induced issues like the consideration of uncertainty or optimisation are more difficult to be addressed.

The characteristic profile of paradigm B (connection of different sub-system simulation models) as shown in Figure 46, reflects the dilemma of those research approaches. Energy and resource flows can be considered on a very detailed level including dynamics and interactions between different sub-systems. This allows

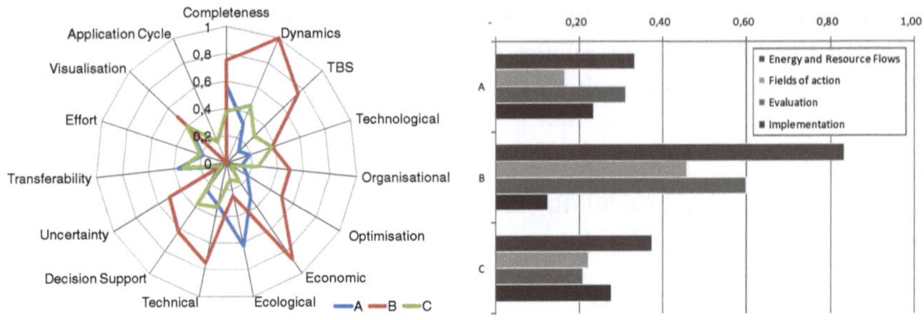

Fig. 46 Criteria fulfilment of energy flow simulation paradigms

covering all fields of actions on a realistic level and complex evaluation schemes including the possibility to consider uncertainty and optimisation. On the other hand the complexity is also the challenge in those cases: different specific simulation models in different expert tools are needed and have to be connected with appropriate dynamic interfaces. This reduces extremely the transferability and increases the effort for both modelling and simulation (e.g. simulation run times, certain computer performance necessary). While being advantageous in theory, a practical application on a broader base does not seem possible yet.

In contrast to paradigm A, paradigm C (direct integration into discrete event material flow simulation) can better depict the dynamics of energy consumption on a system layer. This also enables a quantitatively and qualitatively better consideration of improvement measures at least on the process/machine and manufacturing system level. Additionally this “one-stop-solution” does not need any other tools and allows seamless transfer to other simulation models. However, evidently the user is depending on available functions and restricted by possible limitations of the utilised simulation tool or the possibility to add/change features is necessary. The integration of energy oriented aspects was not in focus of material simulation developers so far. Additionally, the necessary logic to integrate dynamic energy consumption, connected processes like those taking place in technical building services and also appropriate evaluation methods are also not easy to combine with the strong discrete event and material flow oriented perspective of available tools. Each addition to the simulation means certain effort for software implementation. As a result, the considered approaches with this background offer less comprehensive energy flow considerations and evaluation systems as well as no, or less detailed modelling of technical building services.

4.4 Derivation of Research Demand

The research demand can be directly derived from the discussions above. The vision and ultimate objective is to be able to realistically consider all relevant energy and resource flows in the planning and operation of manufacturing systems simultaneous with time, quality and cost related variables. As mentioned, this necessarily involves simulation as well as diverse additional criteria that have to be fulfilled. The analysis of the current state of research clearly reveals that additional work is needed on many aspects:

The **diversity of relevant energy flows in manufacturing systems is not reflected so far**. For obvious reasons (e.g. relevance in industry, easy to measure, direct cost impact) external energy carriers and specifically electricity is mainly focused up to now. However, other important energy carriers need to be integrated in future research. This specifically involves internal energy carriers like compressed air or process heat.

While the state based energy consumption behaviour of single production machines is often addressed sufficiently already, their **interactions and cumulative effects on a system level need to be considered in more detail** as this is the main base for e.g. realistic energy cost calculation as well as planning and control of technical building services.

Technical building services are rarely and qualitatively not sufficiently considered in current approaches. One main reason is that this involves a totally different discipline compared to production engineering. As a result, in production planning there is a lacking awareness of their relevance as well as lacking knowledge about relevant connected energy flows. Additionally, production-planning tools like material flow simulation, currently does not focus, or are even not able to depict those important energy consumers.

Regarding fields of action to improve energy efficiency, **most approaches focus on selected measures and do not provide an integrative view of the whole factory system** although there may be conflicts of goals or rebound effects as well as further potentials for improvement. The possibility of actual optimisation of critical parameters in terms of energy efficiency is not considered at all in most cases.

Besides technical issues, existing energy oriented simulation approaches mostly offer **either an economic or ecological evaluation – but rarely both dimensions simultaneously**. With regard to the demand of increasing sustainability in manufacturing this should be considered in more detail in future research. This also includes more extensive discussions and **integrative evaluation systems to balance possible conflicts of goals between different dimensions**. Additionally, evaluation methods themselves should be more detailed. Specifically the economic evaluation mostly **lacks detailed energy cost models** which are based on energy supply contracts (including e.g. peak surcharges). The **issue of uncertainty** is very important in simulation studies due to stochastic effects that might significantly influence results. This has to be addressed in future research.

At the moment energy oriented simulation is still rather a research topic and less used in industrial practice. Developed solutions are very specific and just transferable with significant effort in terms of time, software costs and necessary expertise. Additionally, the general challenge for material flow simulation is the applicability for flexible manufacturing systems, which is also true here. Altogether there is a **strong need for flexible and easy to use solutions in order to enable broad applicability especially in SME**. Another important weakness until now is the total **lack of an application cycle** that guides the user through the whole process of setting up an appropriate simulation model (e.g. selection of system boundaries, validation) and its parameterisation with measured (consumption) data. Finally, **active decision support** is necessary, which needs a meaningful visualisation of energy and resource flows (above standard material flow visualisation) and connected key figures (e.g. distribution of consumption to different machines and operating states).

All these aspects underline the main challenge which has to be met in context of an energy oriented simulation approach: "complex tasks, simple tools" meaning the necessary consideration of dynamics and complexity of the problem while simultaneously facilitating broad applicability in industry through appropriate design of the simulation environment. In this context the discussion of different implementation paradigms shows possible solutions with their specific strengths and weaknesses. All these considerations will be of main interest when deriving an own energy oriented simulation approach in the following chapters.

Chapter 5
Concept Development

The previous chapters pointed out the necessity and potentials of energy flow oriented manufacturing simulation and showed that there is significant research demand. Against this background the proposed solution approach will be presented in the following chapter. This includes the conceptual development and the detailed description of involved elements.

5.1 Synthesis of Requirements into Concept Specifications

For the development of the proposed solution approach a deductive procedure was chosen. This means that comprehensive general thoughts are brought together in order to form a solution, which is able to address specific problems towards energy efficiency in manufacturing companies. Figure 47 uses the factory life cycle (according to Schenk, 2004) to depict intended fields of application for the concept as a whole. On the one hand, it aims at supporting the development of new facilities (greenfield) through testing of design and control alternatives of the factory system. On the other hand, the concept can also contribute during operation of a factory (brownfield). It allows deriving and evaluating control strategies and also technical and organisational measures, which adapt running production towards more energy efficiency.

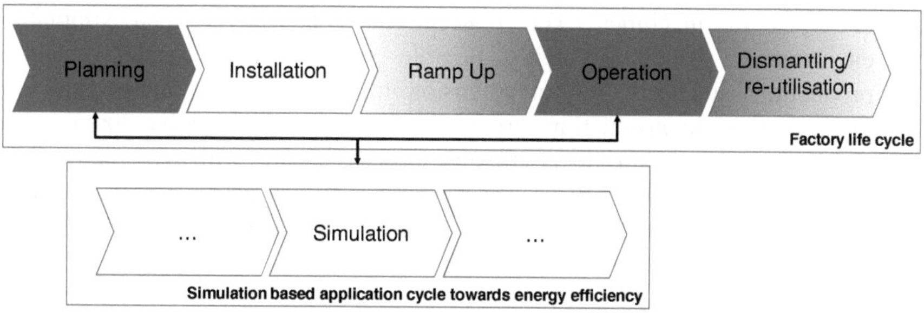

Fig. 47 Classification of proposed concept in factory life cycle according to (Schenk, 2004)

S. Thiede: Energy Efficiency in Manufacturing Systems, SPLCEM, pp. 89–144.
springerlink.com

Based on the necessary scientific-technological theoretical background and industry driven obstacles, diverse requirements for the solution approach were identified. Herewith, energy oriented manufacturing system simulation was identified as a promising approach for improvement. In order to assess the state of research in this field, evaluation criteria were derived through further specification of the requirements (also incorporating additional simulation specific requirements). While further research demand was identified, all these criteria also frame the necessary background and characteristics for the proposed solution approach. On this base the objective is to develop an energy flow oriented manufacturing system simulation approach which

- is **not related or restricted to a specific case but generic in nature and applicable to manifold production situations** in the sense of a generic simulation environment.
- explicitly pursues a holistic perspective including **all relevant energy and resource flows as well as their interdependencies**.
- is **also applicable for small and medium sized enterprises** typically facing obstacles towards energy efficiency measures and usage of simulation.
- is **embedded in a guided methodology** for goal-oriented identification and realistic as well as **multi-dimensional evaluation** of improvement measures in all relevant fields of actions.

With these objectives in mind, Figure 48 shows the mapping of the criteria with actual means as to how they shall be addressed. It can be distinguished between rather general specifications and more specific features envisioned to be integrated. However, all aspects are important and altogether determine the embodiment of the concept as a whole. The specific characteristics are of major importance for understanding the idea of the approach and will be briefly explained in the following. More details to specific aspects can be found in later detailed description of functional principle and modules.

- **Establishment of flexible simulation environment:** It is crucial to understand that in contrast to some approaches presented in the state of research a generically applicable concept is proposed. It is not a single simulation study of a specific company or addressing specific industry sectors or types of machines. The concept aims at providing a simulation environment, which is highly flexible and able to consider manifold production situations. As an extension to available approaches this also includes flexible production structures with no rigid machine linkage and changeable flow of products (flexible process chains with free planning of capacity allocation) throughout the factory. While being reality in many companies (often in SME) these structures are usually hard to model and simulate with conventional tools.

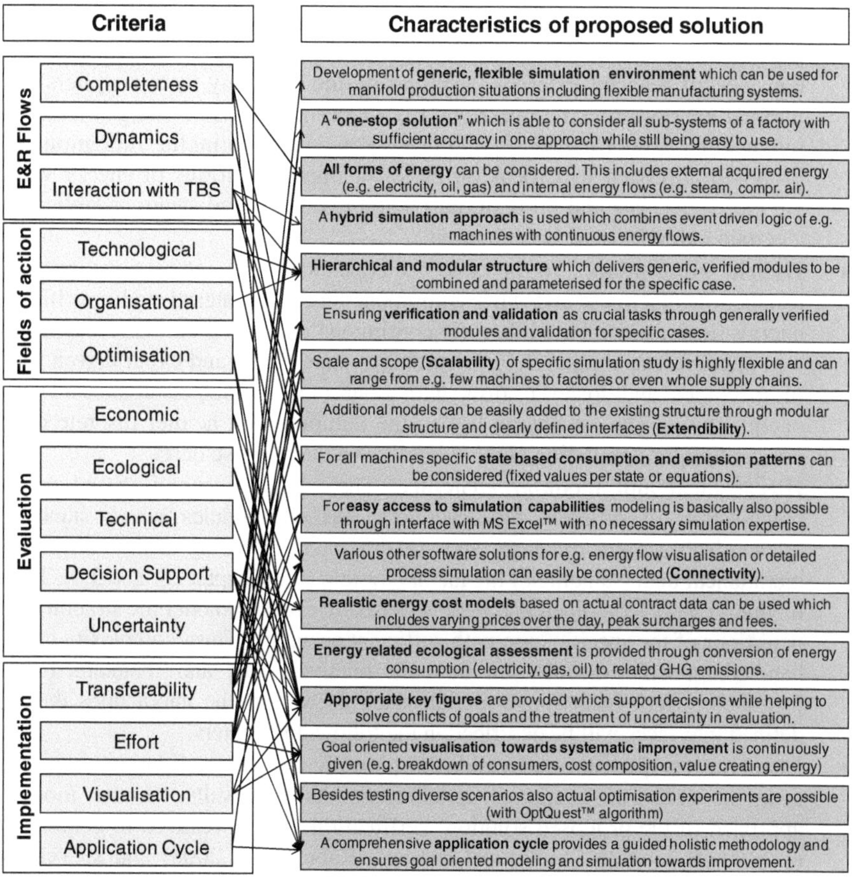

Fig. 48 Mapping of criteria and specific characteristics of the proposed solution

- **One-Stop approach:** The previous chapter revealed different paradigms, which were already applied for energy oriented manufacturing simulation. As described, a combination of detailed simulations of different relevant sub-systems in a factory is a promising approach (paradigm B). This paradigm is able to consider energy flows and interactions realistically and offers a broad range of improvement opportunities. However, resulting complexity and effort is a major disadvantage which impedes implementation. The proposed solution aims at using the advantages while overcoming this drawback. It will be able to simulate not only the manufacturing system itself but also other relevant sub-systems in the factory with sufficient accuracy. However, this will all take place seamlessly within one simulation environment, which significantly facilitates usability, increases transparency of energy flows in the factory system and opens new perspectives towards improvement. Thus, it can be described as a synergetic integration of paradigms B and C (Figure 45).

- **Consideration of all relevant energy inputs:** The proposed solution approach explicitly does not exclude per se any energy flows. It can basically consider all relevant external (acquired by the company from suppliers) and internal energy flows (conversions from external forms of energy) and their interactions (see Chapter 3.2). For reasons of meaningful limitation, the proposed approach will focus on the most important forms of energy which are electricity, gas and oil (external), compressed air and steam as well as (to a certain extend) waste heat (internal).
- **Hybrid simulation:** To be able to cope with the interactions between production machines and TBS equipment through material and specifically energy flows, a combined discrete-continuous or hybrid simulation approach was applied (see chapter 3.4). It combines state based and event driven logic of machines and material flow with continuous variables connected with energy flows. The advantage lies in the combination: neither discrete event nor continuous simulation alone would serve the purpose here.
- **Hierarchical and modular structure:** The whole concept is based on the combination of different pre-structured, verified modules which stand for relevant sub-systems/functions and are structured according to the holistic and hierarchical understanding of the factory system (see Chapter 2.1 and 3.2). These modules can be combined and parameterised in order to model the structure and behaviour of the specific case with sufficient accuracy. Thus, complexity can be handled: specific models can be built relatively fast and maintained with justifiable effort. Each single module itself as well as the interactions through defined interfaces will be described in the following chapters.
- **Ensuring verification and validation:** Verification and validation are crucial tasks in order to guarantee reliable simulation results. In the modular approach of the proposed solution, verification will be realised in general for each single module ("submodel testing", Rabe et al., 2008) and so verified modules will be part of the concept. Validation can just be done in context of the specific case. Thus, it will be part of the suggested application cycle.
- **Scalability:** Scale and scope of the specific simulation study can be adapted according to the actual purpose. Models can be built on a relatively detailed level with consideration of each single manual or automatic process in production. However, if the requirements concerning validity and study objectives are met, it is also possible to leave out irrelevant steps or summarise them into one process. Additionally, there is flexibility regarding system boundaries: studies can either consider a selected process chain of few processes (e.g. for a specific process), a factory as a whole or even complete supply chains (e.g. each company is just depicted as one black-box with inputs and outputs in the simulation).
- **Extendibility:** The modular structure and clearly defined interfaces allows the extension with any additional models. As mentioned, for reasons of meaningful limitation some sub-systems will not be considered in detail here (e.g. air conditioning). However, they can be easily embedded as additional modules at any time. Similarly, other relevant energy or material flows could also be integrated.

- **State based consumption and emission patterns:** As described energy related consumption and emission is not static but depending on the operation mode of the machine (see Chapter 2.3.3). This is addressed through the development of a generic machine state model, which can be parameterised with certain consumption (and other) values or equations for each individual state. As sum of single machine profiles, cumulative load profiles for the whole considered system are also generated during runtime.
- **Easy access to simulation capabilities:** Chapter 3.4 pointed out that the application of simulation techniques in companies may demand for expertise and significant effort in terms of time and connected costs. This often impedes the usage and therewith the utilisation of opportunities towards energy efficiency, specifically in SME. Thus, the proposed solution aims at enabling broad utilisation of simulation capabilities without facing these entry barriers. Therefore two modelling alternatives can be pursued by the user: on one hand the manufacturing system can be modelled directly in the simulation environment as usual. Through graphical modelling and pre-defined modules and interfaces this does also not require extensive simulation experience. On the other hand, through appropriate interfaces the configuration of structure, parameters and production management issues can even be done with familiar software like MS Excel™. In this case, apart from one click for starting the simulation run, users do not have to get in touch at all with the simulation software itself while still be able to tap the advantages of the approach.
- **Connectivity:** At different stages of the proposed solution the usage of additional software tools might be promising for extended applications. This is possible through the open architecture and interfaces. Examples are the aforementioned embedment of office software for model setup, extended evaluation and visualisation opportunities (e.g. energy flow visualisation) or detail simulations of processes or machines (e.g. FEM process simulation).
- **Realistic energy cost models:** Chapter 2.4.1 pointed out that specifically electricity costs are based on individual supply contracts and contain diverse cost components. In order to enable a realistic calculation of costs for the specific case, detailed contract models can be explicitly configured.
- **Simplified energy related ecological assessment:** The proposed approach focuses on energy inputs and less on a comprehensive gathering of all (auxiliary) material flows even though that would be possible. However, this would be necessary for detailed environmental evaluation like LCA (Life Cycle Assessment). In this book ecological evaluation will concentrate on energy related assessment, meaning realistic conversion into the related green house gas (GHG) emissions.
- **Appropriate key figures:** In order to be able to give necessary decision support and solve conflicts of goals, appropriate key figures will be provided. This also includes appropriate statistical treatment in case of probabilistic models (uncertainty). Detailed information will be given later in this chapter.
- **Visualisation towards systematic improvement:** The proposed solution will not only provide cumulative results for general evaluation. It shall also deliver

detailed breakdowns of consumption, environmental impact or cost composition, which help to consciously focus on main drivers.

- **Integration of optimisation possibilities:** Optimisation studies are not in the main focus of this book. However, some possibilities will be considered in the solution approach.
- **Provision of application cycle:** The concept as a whole consists of two main elements. Besides the simulation approach itself, this is a comprehensive application cycle that shall enable implementation in companies. It provides a guided methodology and ensures a systematic modelling and simulation towards system improvement.

5.2 Abstraction of Conceptual Framework

As a first necessary step towards implementation of the concept, an abstraction of the real system into a conceptual modelling framework has to take place. Against the background, the aforementioned theoretical background, Figure 49 shows the control loop of production management extended with the holistic understanding of the factory system, which builds the theoretical backbone of the pursued simulation approach. Based on these considerations five main constituting elements – here called modules - can be distinguished (Figure 49):

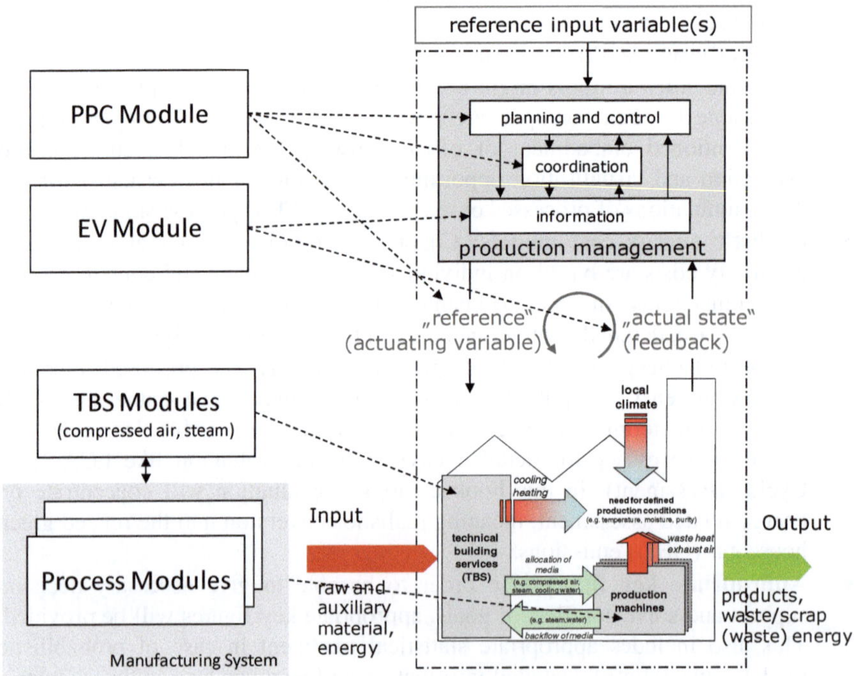

Fig. 49 Contribution of Simulation Modules within Control Loop of Production Management

- Referring to the definitions given before, **Process Modules** (PM, module I in Figure 51) describe the "activities which transform inputs into outputs" (DIN 9000) taking place manually or on production machines. In practice, certain activities [value creating and supporting activities like transformation, combination, transport, control, measure or storage (Barbian, 2005)] are executed simultaneously or successively **directly** with one or more entities (parts) in order to strive towards final products. While being generic in nature, through appropriate parameterisation a process module can imitate the behaviour of any production process and connected machine in terms of process times, failures or energy consumption. Detailed descriptions can be found in Chapter 5.3.1.
- **TBS Modules** (technical building services, (module III in Figure 51) supply the necessary process energy in different forms (Hesselbach et al., 2008b) and therewith enable the operation of process modules. However, while being crucial for running the system they are not interacting with product related entities themselves so their contribution to value creation is rather **indirect**. As described before a broad variety of different types of TBS relevant for energy flows in the factory can be distinguished (e.g. compressed air, steam, cooling, space heat). All these types are based on different physical coherences and work very differently. Therefore, each type of TBS sub-system will be depicted by separate module classes, e.g. there will be general modules for compressed air and steam. However, each of these modules is again generic in nature, meaning that through parameterisation the supply system for the specific case can be considered. Detailed descriptions of implemented TBS module can be found in Chapter 5.3.3 and 5.3.4.
- The generation of products usually involves several process steps, which form process chains (time based logic sequence towards final product) and manufacturing systems (fixed set of available processes and machines used for one or several process chains). Also the supply with different forms of energy usually does not relate to single processes but systems with several consumers. Thus, there is a need for an additional layer that allows the combination of process modules and mutual interfaces with TBS modules. This combinatory layer is realised through the **Manufacturing or Main System Module** (MS, module V in Figure 51, detailed description in Chapter 5.3.7).

Process and TBS modules coordinated through the manufacturing system module constitute the real, "physical" factory system. As described before all these elements strongly interact and altogether determine both production performance and energy consumption of the system as a whole. Single consumption profiles of production machines lead to cumulative load curves for the manufacturing system. TBS-related energy demand of the production equipment (e.g. compressed air) serves as input for appropriate partial TBS-models (e.g. for generation of compressed air). Herewith additional energy consumption (e.g. electricity needed to generate compressed air) of TBS is calculated – together with direct energy

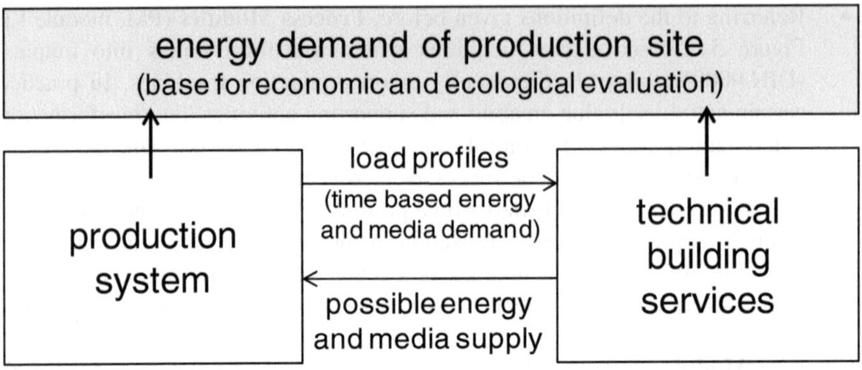

Fig. 50 Simulation based interaction of manufacturing system and technical building services

consumption of production equipment, this leads to the total energy demand of the production site (Figure 50). Additionally TBS models simulate the possible supply with energy or media. Interacting with the manufacturing system a lack of e.g. compressed air (air pressure to low) leads to failures of production machines.

Besides the physical embodiment of the factory system, the production management is considered through different modules. Their allocation within the control loop of production management is also shown in Figure 49.

- The **Evaluation and Visualisation (EV) Modul**e (module IV in Figure 51) gathers necessary data in order to assess the state of the considered system and processes this information in meaningful form as key figures or diagrams (detailed description in Chapter 5.3.6).
- Finally, the **Production planning and control (PPC) Module** (module II in Figure 51) allows the determination of technical and organisational framing conditions for the specific case. This includes the definition of product process chains in general and also their flow in terms of timing and quantity of production machine occupation (detailed description in Chapter 5.3.5).

The diversification of production management into two separate modules was made because two different logic functions can be distinguished. From a technical point of view a separation facilitates modelling as well as verification and validation. In practical application, the connecting element of those modules is basically the user itself. Based on the evaluated results of simulation runs he will develop and adopt production management alternatives, which can be embodied through the PPC module.

As a combination of all thoughts above, Figure 51 shows the conceptual framework of simulation approach as a whole with its five main modules. All modules will be described in detail in the following sections. Afterwards, in order to support the goal-oriented implementation of the simulation approach, a comprehensive application cycle will be developed.

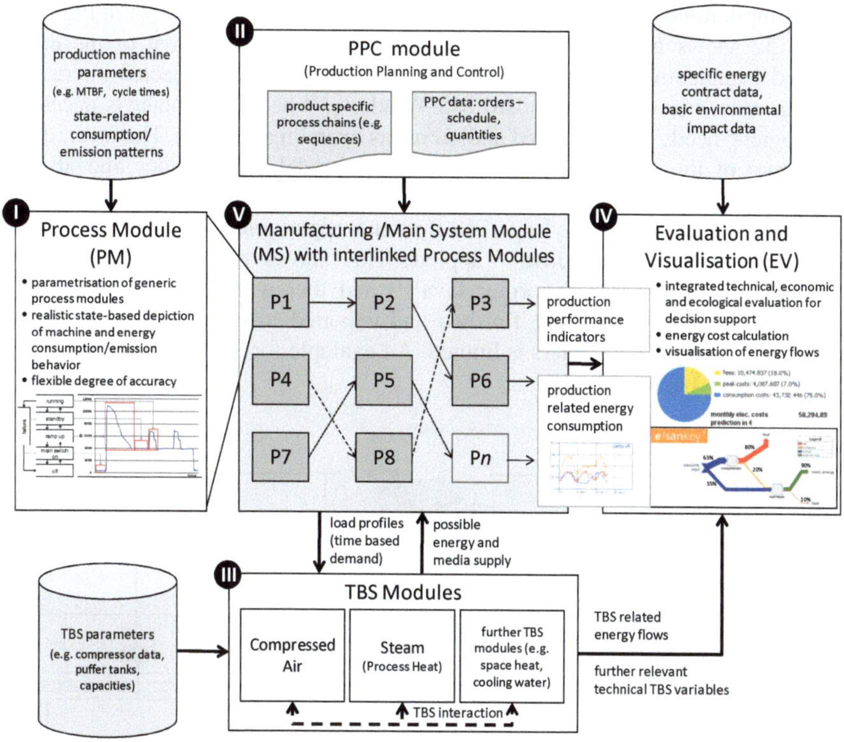

Fig. 51 Conceptual Framework of the proposed simulation approach

5.3 Description of Simulation Approach

5.3.1 Implementation and General Functional Principle

According to the intended specifications and conceptual framework, the simulation environment was implemented in a working solution, the **energy oriented manufacturing system simulation environment**. Besides PM, EV, MS and PPC modules, with compressed air and steam supply two types of TBS modules were already developed, verified and applied to different cases. The proposed simulation environment was embodied using the simulation software tool AnyLogic™ 6.X from XJTek which offers all prerequisites for implementing the characteristics of the solution approach. As one of few tools commercially available, it supports multi-paradigm simulation, which is a basic requirement for many energy flow related features. The development of the innovative simulation solution also demands certain flexibility and freedom which cannot be provided by mature, large scale simulation tools like Plant Simulation™. This is also the reason why those tools were just applied in context of energy simulation through using interfaces with additional external programs. However, this would contradict one of the basic ideas of the proposed solution approach.

Having in mind the conceptual respectively logic framework shown in Figure 51, Figure 52 shows a general overview of the practical embodiment of the energy oriented manufacturing system simulation environment and connected interactions with the user. For usage a standard MS Windows™ based computer is necessary. AnyLogic™ needs to be installed and serves as operating platform. However, due to the nature of the developed approach with its predefined modules and also the possible control via MS Excel™ interface, there is no deep AnyLogic™ expertise necessary. Additionally, AnyLogic™ offers the opportunity to comfortably export models as self-executable Java™ applets. This allows running the model with standard web browsers (supporting Java™) and therewith broadens the field of possible users. However, once finished and exported to the applet, changing the model structure and parameters is limited. An example for an applet based model is given later in Chapter 6.4.

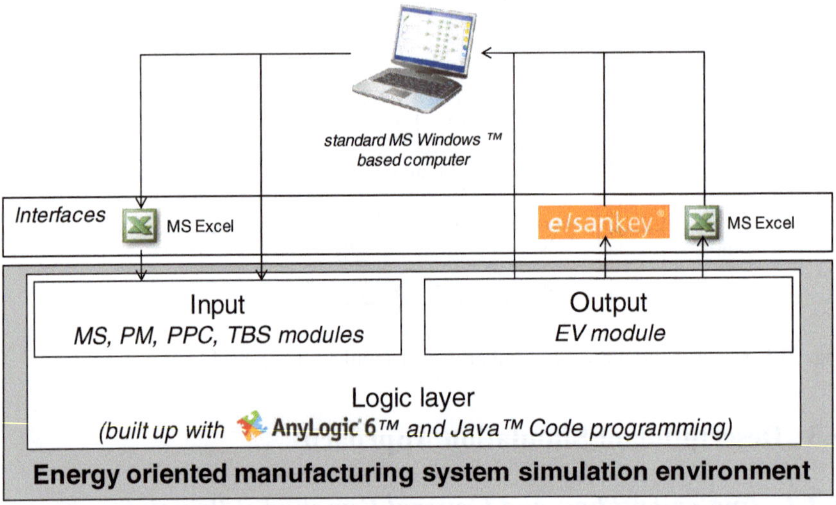

Fig. 52 Practical implementation and user interactions with developed energy oriented manufacturing system simulation environment

The following description of the functional principle concentrates on the more typical case of application with installed AnyLogic software. When starting a new project basic modules PPC, EV and TBS are already given and their interfaces are properly connected. For the **modelling** of the manufacturing system structure generic process modules (PM) are used, which can depict any operations with sufficient accuracy depending on their specific **parameterisation** (a detailed overview regarding available parameters follows in the next chapter). Two alternative ways can be followed at this point:

- For the case of logically and/or technically linked processes respectively process chains, the manufacturing system structure (MS module) can be modelled directly within the simulation tool through drawing the process modules in the main area and linking them. The parameterisation is as well conducted within the tool itself.

- A generic set of process modules can be used and configured through the PPC module, which is based on MS Excel™. Thus, modelling within the simulation tool itself is not necessary.

Also specific **production programs** (sequences, starting times, lot sizes of orders) can be defined within the PPC module. However, simple runs without complex production programs (e.g. only continuous production of certain quantities) can be configured and started independently from this module. When running simulation all relevant variables are transferred to the **evaluation and visualisation** module. Therewith, relevant key figures and meaningful diagrams are generated and displayed continuously during runtime (and can be stored in MS Excel™ based databases). Additionally, energy costs and environmental impact are computed based on parameterised calculation schemes (e.g. electricity contract model). Additionally, energy flow data is continuously transferred to E!Sankey™ which allows a live visualisation in order to depict main energy flows.

In the following sections all five modules will be described. In order to support the understanding of module functionality and interfaces with involved inputs and outputs a standardised illustration will be used (Figure 53). The description of each module will also be conducted in standardised manner. For the core PM and TBS modules the necessary abstraction of the real world system to the functional logic of the module will be explained first. Afterwards, the practical embodiment with involved parameters and state variables will be described and finally a verification study will be conducted for each of those generic modules. MS, PPC and EV modules serve more superior purposes and are responsible for ensuring the general functionalities of the simulation approach. Their necessity and relation to the real system was already pointed out in Chapter 5.2. However, parameters and state variables will also be explained for those modules. While being strongly connected, verification can hardly be conducted independently. Thus, it is realised for MS, PPC and EV module at once in Chapter 5.3.7.

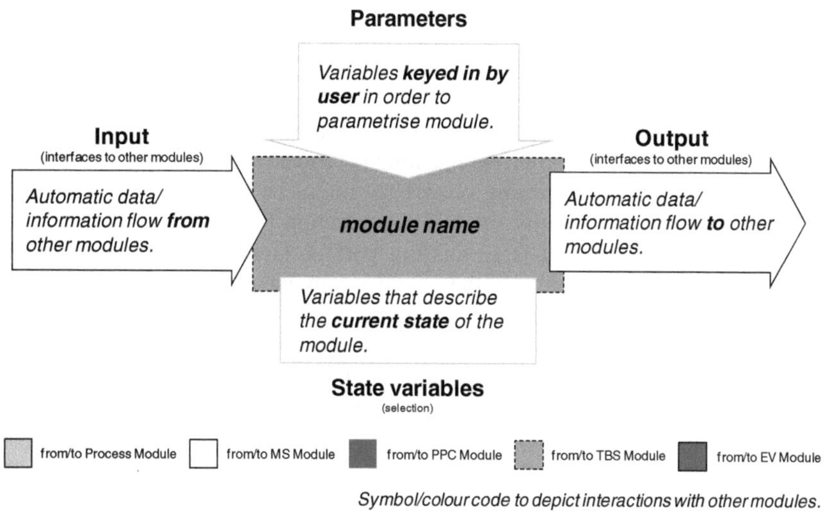

Fig. 53 Description of standardised illustration for modules

5.3.2 *Process Module*

Abstraction from Real World to Model

As described before manufacturing processes are a transformation from inputs to wanted and unwanted outputs. For automated or semi-automated process execution typically production machines are involved. For those machines different distinctive operation states connected through different transition conditions can be distinguished (Dietmair and Verl, 2008; Dahmus and Gutowski, 2004; Devoldere et al., 2007). This well-established perspective on operation states and transitions builds the background for the embodiment of the process module. Figure 54 shows a simplified illustration of the underlying logic of the process module as state chart (in the actual model there are a couple of additional states to ensure functionality etc, which will not be considered here as they are not necessary for the general understanding).

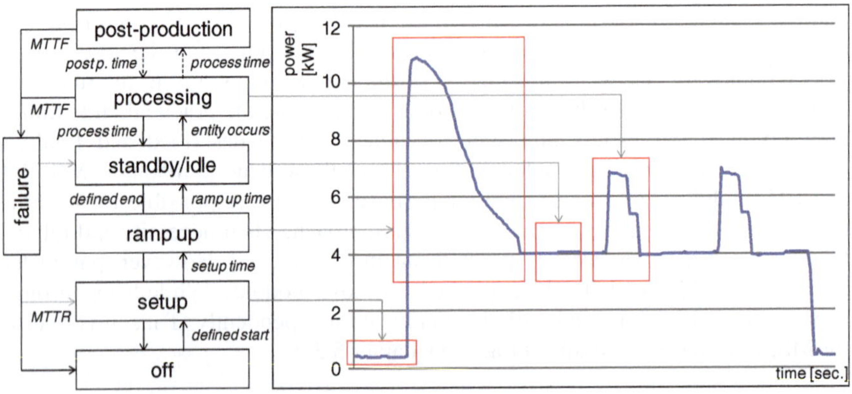

Fig. 54 Underlying state chart logic of process module and connected modelling of (e.g. energy) consumption of machines

Setup is the state when preparing for operation e.g. exchange of tools/dies. *Ramp up* denotes the state after the machine is turned on – thereby all components are enabled and eventual heating phases etc might be necessary. In *standby/idle* mode the machine is basically ready for production but certain requirements for processing are not fulfilled (e.g. missing part or lacking supply with energy). *Operating/processing* stands for the actual treatment of a part. While being an abstracted consideration of manufacturing processes here, it is less relevant what physically happens during the process. In the end a manufacturing process somehow transforms one or more entities at once, which consumes time and additional inputs like e.g. energy. This necessary abstraction allows a very broad applicability to any manufacturing processes. The state *post-production* was introduced because machines may include certain automatic procedures after processing (e.g. cleaning). As displayed in Figure 54, the transition between different states can be triggered by either time or occurrence of events:

- **time:** pre-defined duration of states depending on machine and product-specific process characteristics.
- **events:** stochastic failures, control signals from e.g. PPC module (e.g. shift start), occurrence of parts in order to be processed.

As formerly mentioned, the *failure* state is of specific importance in context of the time based behaviour of the process module. This state is determined by two characteristic variables, which can be derived from company data: Mean Time To Failure (MTTF) is an expression for the average lifetime and can serve as parameter for describing the probability of a failure. Mean Time To Repair (MTTR) determines the duration of a failure (Bertsche, 2004). For mathematically describing the failure behaviour, distribution functions (e.g. normal, exponential) are used. For practical purposes and its broad applicability the usage of the Weibull function is suggested (Abernethy, 2006; Bertsche, 2004; Birolini, 2010).

$$f(t) = \frac{dF(t)}{dt} = \frac{b}{a} * \left(\frac{t}{a}\right)^{b-1} e^{-\left(\frac{t}{a}\right)^{b}} \tag{19}$$

The necessary input parameters b (shape parameter, describes the mode of failure) and a (scale parameter, characteristic life time at which 63.2% of considered units fail) can be derived as a result of statistical analysis of e.g. process times or failure occurrence and transformation through mathematical/statistical methods like the Median Rank Regression (MRR) or Maximum Likelihood Method (MLE). The Weibull function is very useful due to its flexibility: all kinds of different distributions can be described with this equation through varying the shape and scale parameters (Figure 55) (Bertsche, 2004; Birolini, 2010). When using the Weibull distribution the determined MTTF is transferred to the scale parameter for a while using the gamma function Γ:

$$a = \frac{MTTF}{\Gamma\left(\frac{1}{b}+1\right)} \tag{20}$$

with

$$\Gamma\left(n_\Gamma\right) = \int_0^{+\infty} e^{-x} * x^{n_\Gamma - 1} dx \tag{21}$$

or (for positive integer values)

$$\Gamma(n_\Gamma) = (n_\Gamma - 1)! \tag{22}$$

As the equations show MTTF will be equal to a if b=1 ($\Gamma(2)=1$), which depicts a random failure behaviour. It is crucial to note that the usage of distribution function automatically leads to a probabilistic nature of the simulation model and any results are only one possible outcome of the simulation. Hence, sufficient simulation runs and appropriate statistical treatment are necessary. Further explanations in this context can be found later in Chapter 5.3.6.

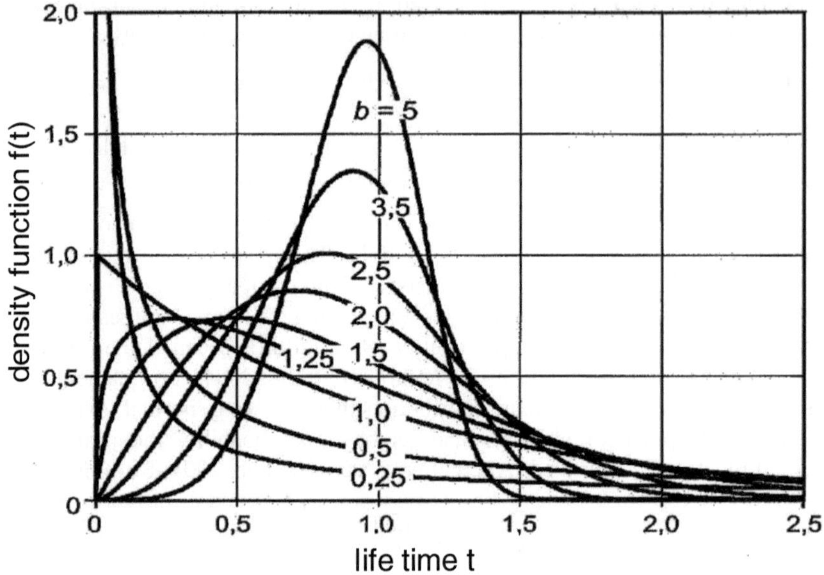

Fig. 55 Weibull function with different shape parameters b (Bertsche, 2004)

As shown in Figure 54 each state is connected with the consumption of energy or/and auxiliary materials. Any kind of consumption can be allocated to different states with either single value for each state (sufficient for many applications) or based on equations. An equation based formulation might make sense if further coherences shall be included. For instance, bearing in mind the thoughts from Chapter 2.3.3 increasing load or machine speed usually lead to higher demand on energy.

Practical Embodiment

Against the background of the conceptual thoughts of the process module, Figure 56 shows an illustration of its main elements and interfaces. In accordance with the definitions in Chapter 2.1 a process involves inputs and outputs which are being determined by the specific parameters. Certain variables can be used to assess the state of the process.

As already mentioned in context of the general conceptual structure, process modules are embedded within the main (MS) module. Thus, technically they do not have direct interfaces with other modules. In terms of input, process modules are triggered by the arrival of one or several entities (parts), which have to be processed. This is determined through production planning and control as well as the processes in the upstream process chain. Additionally the operation of process modules depends on energy supply, e.g. a certain compressed air pressure or the availability of necessary steam. Although provided by TBS modules, technically this information is as well transferred to each process module via MS module.

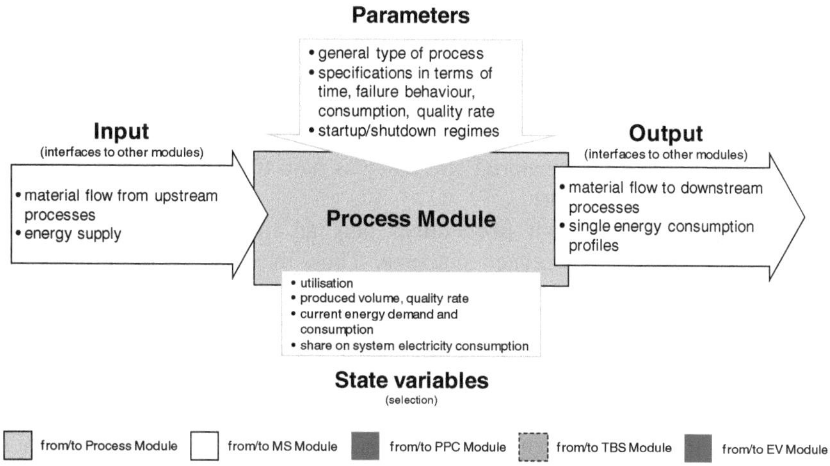

Fig. 56 Constituting factors of Process Module

Parameters

A summary of relevant parameters is also shown in Figure 56. Due to their importance Table 18 gives a detailed overview of all parameters including their actual identifier within the simulation, a short description as well as the unit and the range of values. In general three different types of parameters can be distinguished for the process module:

- As a very basic option the **type of process** needs to be defined. As mentioned before, the simulation does not aim at a detailed representation of any processes. However, two basic alternatives are distinguished, which can describe the generic logic of most processes relevant in manufacturing. If the *transportation* parameter is enabled, the part is constantly moving with a certain speed and for a specific distance (together this determines the duration) and leaves the process module afterwards. A typical application would be e.g. a conveyor belt. Alternatively (*transportation* parameter off), each single part (respectively a lot of parts) is delayed for a certain period of time without movement where normally the value creating processing takes place. In contrast to transportation activities, other parts cannot be processed while the process is occupied. This is the case for many manufacturing processes like e.g. metal machining (machine tools).
- In the next step, **individual technical specifications** need to be inserted in order to enable realistic behaviour. Thereby the described state based perspective on processes and machines is relevant to keep in mind. The duration and related energy consumption for each state can be keyed in with fixed values or also based on equations/distribution functions. Failure behaviour can be configured through applying the Weibull function. Practically, the standard syntax (provided by AnyLogic) *weibull(shape*

parameter, scale parameter, minimum value) needs to be used here. For instance, a normal distribution (around the value of e.g. 500 seconds) can be realised with the term *weibull(3.4, 500, 0)*. Besides state based characteristics further general or more product oriented technical specifications can be included. Examples are the necessary compressed air pressure for operation, the quality rate or the material efficiency as ratio from weight of finished part to input material per part.

- **Startup and shutdown procedures** play an important role for machine operation and energy related outcome. Thus, they can be defined for each machine individually and different strategies can be integrated.
  - The individual machine can start – which means turning it on, not necessarily processing - at a definable point of time during the workday, usually at the beginning of the shift (e.g. 6am). The alternative is to turn it on when the first order appears.
  - The machine can stop operation at a specific point of time, e.g. at the end of the shift.
  - Even more, an automatic shutdown can be enabled. Therewith, the machine shuts down automatically if idle for a definable period of time.

Table 18 Parameter list of process module

Parameter	Description	Dimension
Name	Labelling of process module	String variable, free naming
process_time	Necessary time per product if dynamic process time is off.	in [sec.]
dynamic_processtime	Process time is not fixed but individual per product.	in [sec.]
set_up_time	Time for setup of machine.	in [sec.]
ramp_up_time	Time for rampup of machine.	in [sec.]
post_production_time	Time necessary for post production processes.	in [sec.]
start_machine_regime	Process can either start at fixed hour of the day or when first order occurs.	1: Machine start if first parts occur (starting with setup) 2: Machine start at specific hour
start_machine_time	Denotes starting hour, e.g. shift start (if start_machine_regime=2)	[hour of the day], from 0-24
end_machine_time	Denotes machine stop hour, e.g. shift end (if start_machine_regime=2)	[hour of the day], from 0-24
lot_size	Number of parts being processed at once.	[number of parts]

Table 18 (*continued*)

Parameter	Description	Dimension
quality_rate	Average percentage of good to total parts.	[from 0..1 meaning 0-..100%]
MTTF	Mean Time To Failure.	[in sec.] plus distribution function, free range
MTTR	Mean Time To Repair.	[in sec.] plus distribution function, free range
transportation	Denotes whether it is a transportation process.	[1: yes, 0: no]
conveyor_length	Length of conveyor (if transportation on)	[in meter]
conveyor_speed	Speed of conveyor (if transportation on)	[in meter/second]
accumulating	Describes whether several entities can use conveyor simultaneously.	[1: yes, 0: no]
material_efficiency	Ratio of product weight and raw material input.	[from 0..1 meaning 0-..100%]
nec_compair_pressure	Minimal compressed air pressure necessary for operation.	[in bar]
power_consumption_*state* *(for setup, ramp_up, standby, producing, post_production)*	State based allocation of electrical power demand.	[in kW]
compair_consumption_*state* *(for setup, ramp_up, standby, producing, post_production)*	State based allocation of compressed air demand.	[in m^3/h]
steam_consumption_*state* *(for setup, ramp_up, standby, producing, post_production)*	State based allocation of steam demand.	[in kg/h]
gas_consumption_*state* *(for standby, producing)*	State based allocation of gas demand.	[in kW]
aux_consumption_*state* *(for setup, ramp_up, standby, producing, post_production)*	State based allocation of selected auxiliary material demand.	[any continuous unit]
auto_shutdown	Activation of auto-off function for machine.	[1: yes, 0: no]
auto_off_time	Time until machine turns automatically off (if enabled).	in [sec.]

State Variables

The state of each process module can be assessed continuously during simulation run time through different variables. An overview of considered state variables and the graphical depiction of a process module during simulation gives Figure 57. The current state is depicted with a colour scheme and with one glance the operation mode of each machine can be recognised. Two variables are of specific interest to evaluate the performance of each process. On the one hand, the utilisation rate is used which gives a clear indication of the process' contribution towards value creation and the employment of provided capacity.

$$utilisation = \frac{value\ creating\ time}{total\ time} \tag{23}$$

On the other hand, the current energy demand of each process is displayed. Even more, instead of just displaying the energy consumption continuously – which would not deliver meaningful information when shown alone – the share of the single process consumption related to the total energy consumption of the system is displayed. Because of its importance electricity consumption is specifically addressed.

$$share\ on\ total\ elec.consumption = \frac{process\ elec.consumption}{system\ elec.consumption} \tag{24}$$

With this key figure hotspots regarding electricity consumption can be identified at a glance. This supports the systematic improvement process towards energy efficiency in manufacturing.

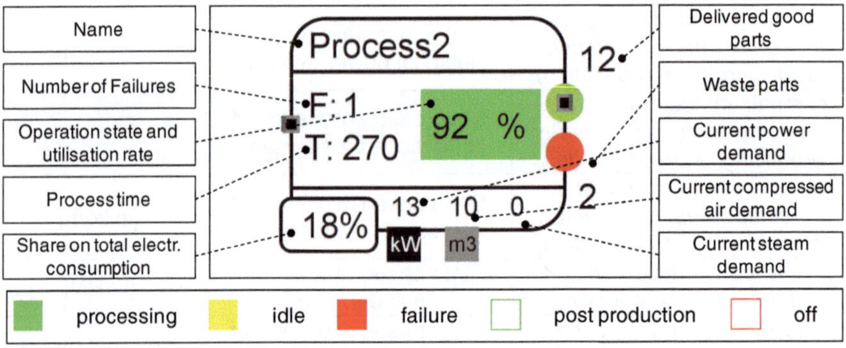

Fig. 57 Screenshot of graphical depiction of process module in simulation

Verification

For verification of the process module a simple case was built up, which can be modelled and simulated but also statically calculated. This is possible as the case is deterministic in nature – no probabilistic/stochastic elements and just one process with defined product arrival are considered. This enables the comparison of two separate models - static calculation and simulation - to describe machine behaviour and calculate energy consumption. If both models come to the same results independently from each other the process module can be considered as verified since its "behaviour is basically consistent with other models" (Chapter 3.4 based on Rabe et al., 2008). Obviously typical simulation applications will be more complex and not solvable with static calculation (which reasons the usage of simulation). However, basic functionalities of the process module itself remain the same and so verification is possible with the simplified case.

The parameters are shown in Figure 58. For this example, two parts of any kind are processed and their appearance at this generic machine is determined at 60 seconds (part 1) and 400 seconds (part 2). Figure 58 shows the simulation results in terms of state variables at the end of the simulation run (1000 seconds). Additionally, the load curve for electricity and compressed air consumption is shown. The first analysis underlines that the simulation model works as expected: the machine starts with the arrival of the first part at 60 seconds and is processing the part after a 30 second ramp-up phase. After some time of idleness at 400 seconds the second part appears and is started to be processed. As intended there is a failure within the operation which leads to stoppage of 120 seconds (MTTR). The machine is restarted after that and is in idle mode until the enabled AutoOff function shuts down the machine automatically after 300 seconds. For this deterministic case, this machine behaviour could be easily predicted with the data given. Therewith, the basic functionality of the process module is verified.

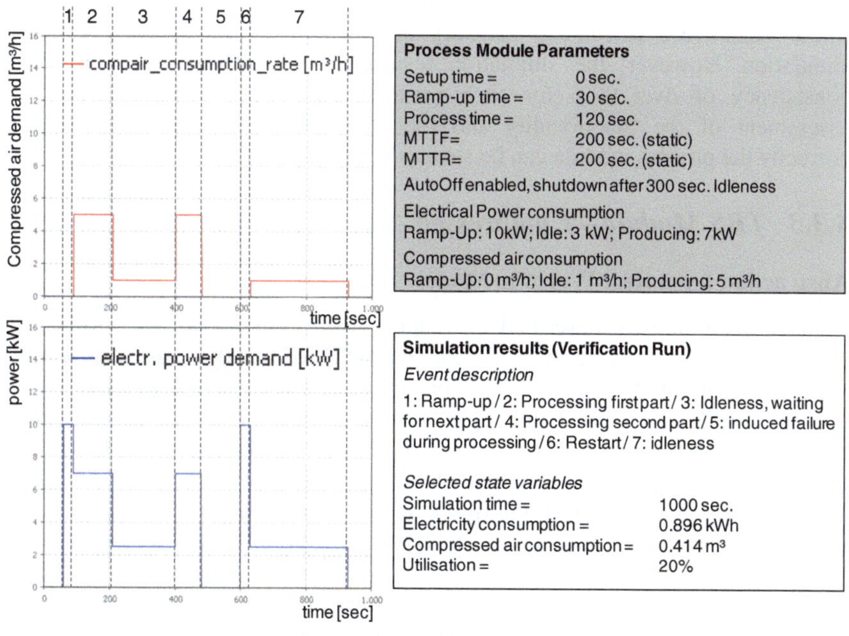

Fig. 58 Results of verification run for process module

Additionally, the means for calculation shall be tested as well. The consumption W can be calculated mathematically through the following equation which combines the multiplied power (P) and time (t) for different states of the machine (ramp-up, idle, process).

$$W = \sum P_{ramp-up} * t_{ramp-up} + \sum P_{idle} * t_{idle} + \sum P_{process} * t_{process} \qquad (25)$$

As mentioned, in this simple case with no probabilistic behaviour and only a single process, all events can be determined in advance. Thus, it is possible to calculate statically:

$$\begin{aligned} W = 30s * 10kW + 120s * 7kW + 200s * 2.5kW + 80s * 7kW \\ + 30s * 10kW + 300s * 2.5kW = 3250\ kW * s \\ = \mathbf{0.903\ kWh} \end{aligned}$$

With the same equation also the compressed air consumption can be determined with:

$$\begin{aligned} W = 120s * 5m^3/h + 200s * 1m^3/h + 80s * 5m^3/h + 300s \\ * 1m^3/h = 1500\ m^3/h * s = \mathbf{0.417\ m^3} \end{aligned}$$

The utilisation of a production machine can be calculated through:

$$utilisation = \frac{value\ creating\ time}{total\ time} = \frac{120s + 80s}{1000s} = \mathbf{20\%}$$

These results are calculated statically and completely independent from the simulation. However, the simulation results in very similar numbers with a consistency of over 99% for each variable. Altogether, with the successful assessment of the functionality and the proof that calculations are working correctly the process module can be seen as verified.

5.3.3 TBS Module – Compressed Air

Abstraction from Real World to Model

Chapter 2.4.2 already described the supply with compressed air with one or several compressors being the main component(s) for transforming electrical into mechanical energy. Determining variables of a specific compressor in this context are the possible compressed air supply (volume) rate, the electrical power demand and the control scheme (Bierbaum and Hütter, 2004; Ruppelt, 2003). Four different control approaches are available in industrial practice, which are shown in Figure 59. As illustrated in the figure, these control regimes directly determine the compressed air output and the energy consumption of the compressor. The actual compressed air pressure in the system is the reference variable for compressor control while keeping up a defined pressure to ensure operation of production machines. Thus, main influencing variables are the different possible operation states (running, idle, off) as well as values for startup and shutdown pressure of the individual compressor.

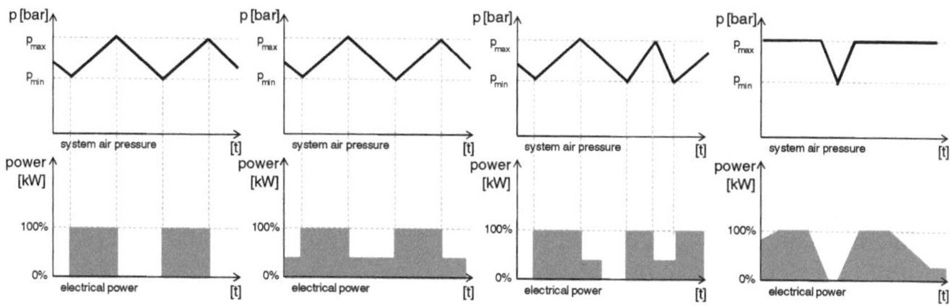

Fig. 59 Integrated control schemes for compressors (Bierbaum and Hütter, 2004)

Orienting on the nominal pressure, the operating compressed air pressure is the result of supply and demand in the system. Based on the equations given earlier this actual compressed air pressure can be calculated with

$$p_{actual} = \frac{V_{supply} - V_{demand}}{V_{system}} + p_{nominal} \tag{26}$$

This inherent logic for determining compressed air system pressure and controlling a compressor builds the background for the compressed air module. As shown in Figure 60, the compressed air supply rate and electrical power demand is modelled based on the distinctive states of a compressor.

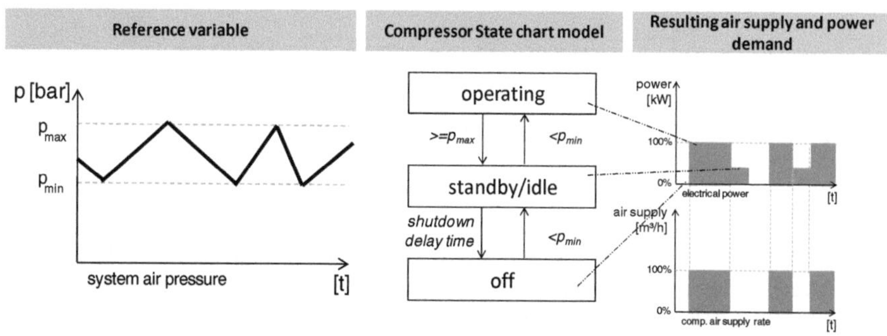

Fig. 60 State based control of compressor in compressed air module

A compressed air system as a whole can contain several compressors in order to provide the necessary capacity but also flexibility for air supply. For control of multi-compressor systems different control schemes are available. For many years, a cascading approach was mainly applied. Thereby each compressor is responsible for a certain pressure range and all work together with cascading overlapping. Nowadays all compressors usually work in one shared pressure range and are controlled electronically (Ruppelt, 2003). Both control schemes can be realised in the module through appropriate parameterisation.

Practical Embodiment

Figure 61 shows the main elements and interfaces of this module. The main input is the total compressed air demand of the system. Based on this demand and the calculated supply the resulting operating pressure is determined and handed over to the MS module from where it serves as input for process modules. Additionally, the compressor induced electricity consumption flows into the EV module in order to calculate the total electrical power demand.[1] Parameters and relevant state variables are described in the following.

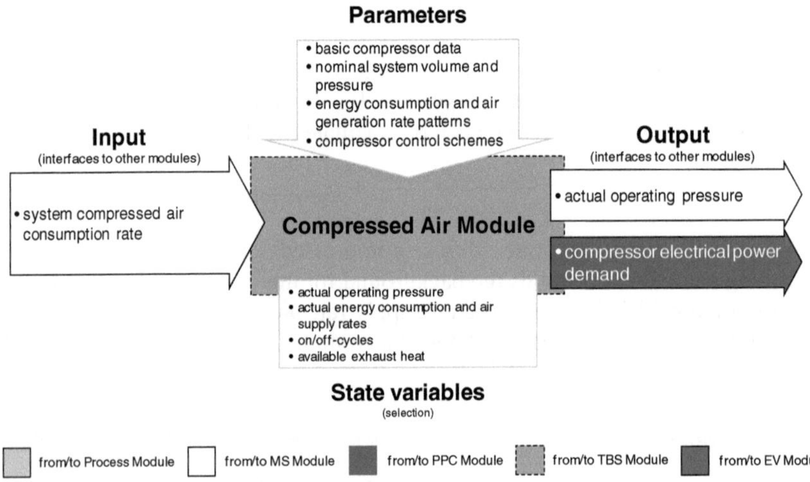

Fig. 61 Inputs, Outputs and Parameters of the Compressed Air Module

Parameter

The parameterisation can be done in either the simulation environment itself or via MS Excel™ user interface which is provided as well. All parameters are listed in Table 19. As explained above (Chapter 2.4.2) the compressed air systems consists of one or several compressors and auxiliary components (e.g. filter), which all need electricity for operation. In the simulation, a compressed park with up to seven compressors can be modelled. For each compressor relevant technical data

[1] The compressed air module can also be used as a separate application (without connection to other modules). In this case, a measured compressed air load profile of the factory is imported as MS Excel™ document. Based on this data the compressed air module calculates the state variables of the system (e.g. electrical power demand, air pressure) with its specific characteristics. Without the need of building up a comprehensive model of the whole factory with all production processes, this allows the analysis of different compressed air system configurations against the background of a given demand of compressed air over time. The consideration of dynamic interactions between production and compressed air supply is naturally limited in this application case.

Table 19 Parameter list of compressed air module (*n*: number of compressor)

Parameter	Description	Dimension
identifier_compressor*n*	Labelling of compressor 1..n.	String variable, free naming
leakage_rate	Air losses in system due to leakage.	[m³/h]
plan_operating_pressure	Nominal air pressure of system.	[bar]
plan_total_volume_compressed_air	Nominal air volume of system.	[m³]
running_power_rate_compressor*n*	Power demand for running compressor.	[kW]
idle_power_rate_compressor*n*	Power demand for idling compressor.	[kW]
auxiliary_energy_consumption_rate	Power demand for auxiliary components.	[kW]
startup_pressure_compressor*n*	Trigger pressure for compressor startup.	[bar]
shutdown_pressure_compressor*n*	Trigger pressure for compressor shutdown	[bar]
shutdown_delay_compressor*n*	Delay time before shutdown from idle.	[seconds]
control_scheme_compressor*n*	Control regime for single compressor.	(1) interrupting control, (2) idle run control, (3) delayed interruption, (4) continuous control
air_generation_rate_compressor*n*	Possible air supply of compressor.	[m³/h]
compressor_efficiency	Efficiency compressor energy conversion.	[%]
heat_exchanger_efficiency	Efficiency of heat exchanger.	[%]

like the electricity consumption rate for different operation state, the possible air generation rate and configuring variables for the underlying control scheme (e.g. type, startup/shutdown pressure) can be keyed in. Auxiliary system electrical power demand can also be added. From a system perspective the nominal air pressure and system volume (as a result of tanks and pipes) are crucial variables to calculate the actual air pressure in the system.

Additionally efficiency factors for the compressor itself and a possible heat exchanger can be defined which are typically given in technical documentation. These values allow further considerations in combination with other modules.

State Variables

Different variables are used to track the state of the compressed air system. Figure 62 shows a screenshot of the graphical user interface of the compressed air module which gives an overview of relevant state variables.

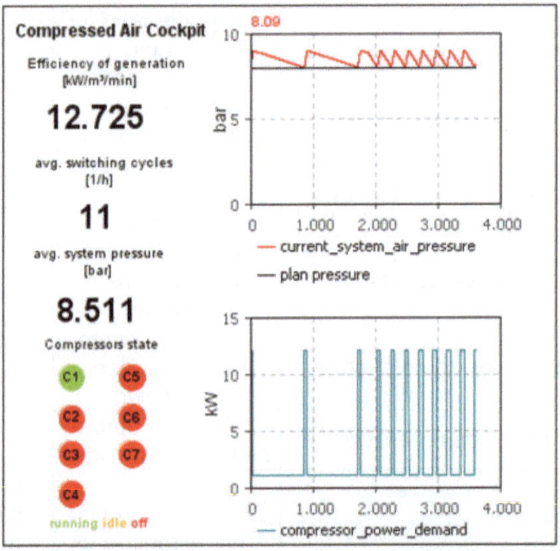

Fig. 62 Overview of relevant compressor state variables (screenshot from GUI of compressed air module)

Of main relevance for the whole factory is the current operating compressed air pressure – as explained above it is calculated based on the compressed air system characteristics and the total air supply and demand of the system. Additionally, the electrical power demand is an important state variable and can have significant influence on total power demand of the factory. It is the result of the electrical power demand of each compressor (depending on specific operating state) and the auxiliary equipment (e.g. filter). Based on the electricity consumption of the compressor system and the supply with compressed air, the specific compressor power demand (kWh/m³/min) related to the classification shown in Figure 23 is continuously calculated. From the technical perspective on compressors with the number switching operations another variable is very relevant: constantly turning on and off the compressor leads to deterioration and reduces lifetime. Therefore, there are certain limits for switching operations, which should be respected (Figure 63).

compressor power [kW]		
from	to	max. switching cycles allowed [1/h]
4	7,5	30
11	22	25
30	55	20
65	90	15
110	160	10
200	250	5

Fig. 63 Allowed switching operations for compressors (Müller et al., 2009)

As mentioned above compressed air involves significant energy losses in form of exhaust heat. Therefore, occurring heat is also an important state variable since it could be used for energy recovery (e.g. for steam generation).

Verification

For verification, data from a literature case study from (Bierbaum and Hütter, 2004, p. 146) is used. Thereby, constant compressed air consumption is assumed which depicts leakage losses in the network. The respective key figures are shown in Figure 64. If the simulation model would be able to come to the same results its functionality can be considered as verified. Energy consumption values were not available but are assumed to be 30kW (operating) respectively 10kW (idle) in order to verify consumption calculation as well. Additionally, a control scheme with delayed (60 seconds) compressor shutdown was selected. Therewith, based on the calculated values of the given case study the energy consumption can be calculated:

$$W_{hour} = 8 * (4.56min * 30kW + 1min * 10kW) = \mathbf{19.57\ kWh}$$

Figure 64 shows the simulation results. As the diagrams show, the model of the compressed air system acts as intended and expected. Also the simulated state variables are very similar to the calculated results provided by the external case study. The accuracy through all variables is at least 98%. This is also true for the electricity consumption, which reflects the statically calculated value very well. Similar to the verification of the process module this is certainly a relatively simple case with constant compressed air consumption. However, only this simplicity enables a correct verification; the general logic stays the same anyway, even for more complex cases with fluctuating consumption patterns.

Altogether, with the successful assessment of the functionality and the proof that calculations are working correctly, also the compressed air module can be seen as successfully verified.

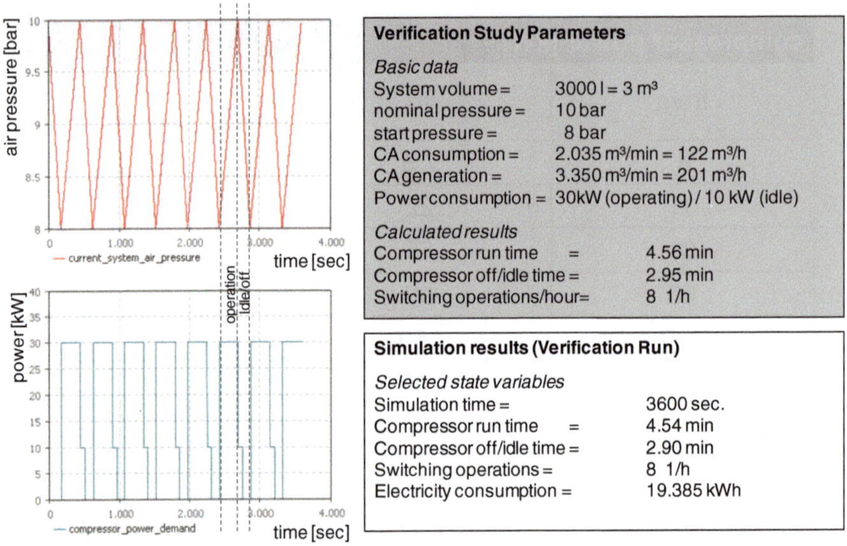

Fig. 64 Verification study for compressed air module

5.3.4 TBS Module – Steam Generation

Abstraction from Real World to Model

The descriptions in Chapter 2.4.3 underlined that, from a combustion perspective, the steam generation process is determined by boiler capacity and efficiency as well as the quantity and type of fuel. With respect to the medium water its specific enthalpy for evaporation (depending on pressure and temperature, a steam table can be found in the appendix) and the share and temperature of both fresh and condensed water are relevant. The underlying understanding of the steam supply module with the considered parameters is depicted in Figure 65.

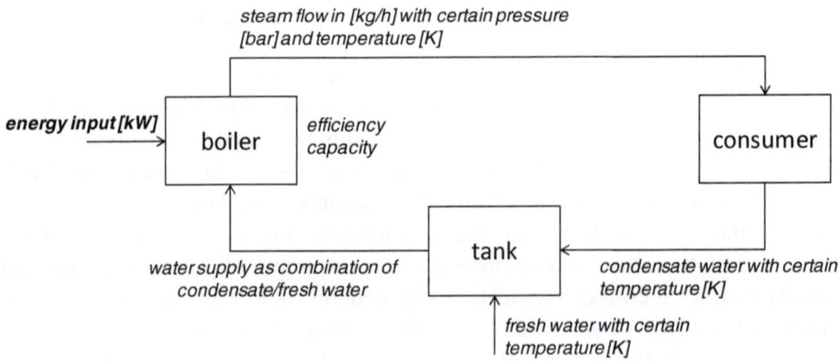

Fig. 65 Abstraction of steam supply system as underlying model logic

The main purpose is to calculate the necessary energy input in kW (and finally the necessary quantity of fuel) based on the steam supply ($\dot{m}_S$, constant flow or dynamic but limited by capacity of boiler). With combining the variables described above (share of fresh water/condensate $n_{FW/C}$, the related temperature differences to the steam ΔT, water specific variables c_m and h_w) the necessary energy input E_S can be calculated with

$$E_S = \frac{\dot{m}_S * [(n_{Fw} * \Delta T_{Fw} + n_C \Delta T_C) * c_m + h_W]}{3600\ kJ/\ kWh} \tag{27}$$

Through additionally applying the equations given in Chapter 2.4.3 the time based energy and finally fuel demand can be calculated. Steam generation is a continuous process that is well describable through thermodynamic equations. Thus, these equations build the inherent logic of the steam module as continuous simulation model. The configuration of the specific system can be modelled with sufficient flexibility and accuracy.

Practical Embodiment

Figure 66 shows the main elements and interfaces of the steam module. The module can use the steam demand of the production equipment as dynamic requirement for steam generation. Alternatively, a constant generation volume can be determined. Based on the specific boiler data, technical specifications of the evaporation process and water supply, the necessary energy input and actual steam generation rate is calculated. This data is handed over to the MS and EV module. As further option, data on available and usable exhaust heat from the compressor system can serve as input of the steam module. With this heat recovery option the demand on oil or gas for steam generation can be reduced.

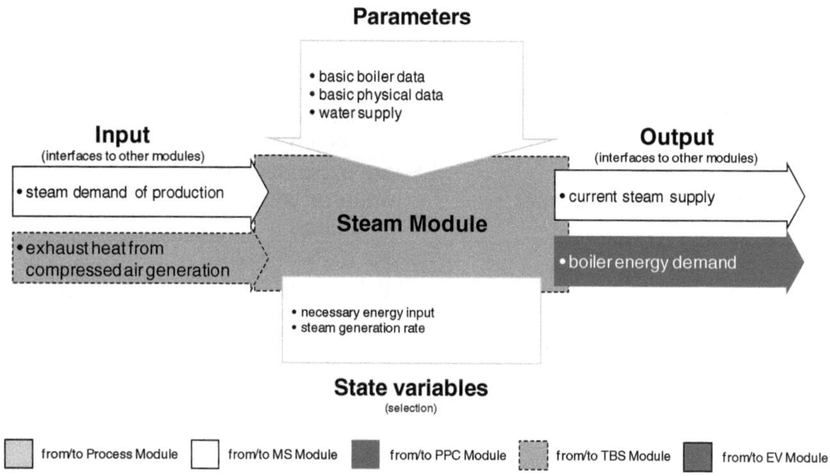

Fig. 66 Inputs, Outputs and Parameters of the Steam Module

Parameters

Table 20 shows a list of all parameters of the steam module. It can be distinguished between

- basic technical specifications of the boiler and control strategy applied in the specific system (efficiency, maximum steam generation),
- information about the evaporation process itself like the targeted temperature or specifications of water supply,
- general physical variables like the specific heat capacity or the evaporation heat, which is depending on the pressure and temperature of the steam and can be derived from steam tables.

These specifications are widely used for dimensioning steam generation system and can be found in technical documentation of the boiler or basic literature (e.g. steam tables).

Table 20 Parameter list of steam module

Parameter	Description	Dimension
boiler_efficiency	Efficiency of boiler for steam generation.	[%]
max_steam_generation_rate	Maximum supply with steam per hour.	[kg/h]
dynamic_supply_control	Steam supply depending on demand but limited to maximum rate.	1: on / 0: off
constant_supply_rate	Constant supply rate with steam per hour (if dynamic control is off).	[kg/h]
target_temperature	Designated temperature of steam.	[K]
share_condensedwater	Share of condensed water on water supply.	[%]
share_freshwater	Share of fresh water on water supply.	[%]
temperature_condensedwater	Temperature of condensed water backflow.	[K]
temperature_freshwater	Temperature of fresh water.	[K]
heat_capacity	Specific heat capacity of used medium	[kJ/kg K], for water 4.187 kJ/kg K
steaming_energy	Specific evaporation heat.	[kJ/kg]

State Variables

Two variables are of main interest for the steam module, the steam generation rate and the necessary energy input. The steam generation rate is being determined by the control strategy applied. Steam generation can either be a constant flow based on a fixed generation rate or connected with production demand (with certain minimum generation rate and limited by maximum capacity of the boiler). Based on this steam generation and the relevant technical/physical variables of the system the necessary energy input is being calculated.

Verification

For verification, a case study with specific numbers is applied. In contrast to the previous modules the steam modules is mainly based on equations. Thus, only the correct calculation of values needs to be verified. Therefore, a static calculation is compared with the simulation outcome (Figure 67).

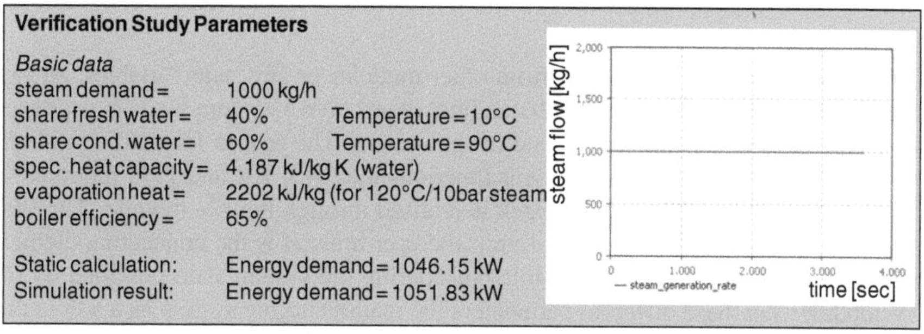

Fig. 67 Verification results for steam module

For many applications a constant supply with heat is necessary; therefore this case is being considered with a constant demand of 1000 kg steam per hour. The additional relevant background data is shown in Figure 67. These values were also used in the simulation model. As Figure 67 shows, the simulation behaves correctly and comes to the same results. Thus, calculation is working properly and the steam module can be considered as verified.

5.3.5 PPC Module

Figure 68 gives an overview over the production planning and control (PPC) module. It shows the specific role of this module as superior layer which is responsible for coordinating all production activities in the manufacturing system.

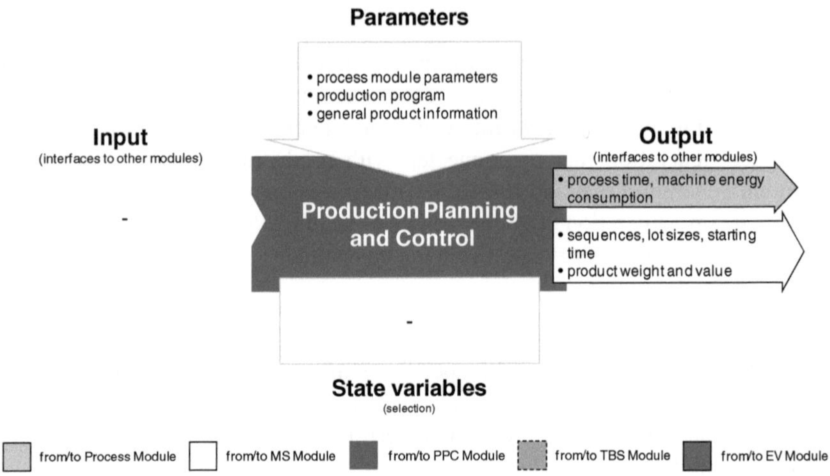

Fig. 68 Inputs, Outputs and Parameters of PPC Module

The PPC module has no input from other modules or any state variables. At first this seems to contradict the understanding of a control system with a necessary feedback variable regarding the state of the system. The reason for that is the logic separation of production management functions into MS, EV and PPC module. As described in Chapter 5.2, the feedback is realised through the EV module. The PPC module is not an automated control unit; the user himself is the connecting element and changes parameters in PPC while considering the system state provided by the EV module. Via these different parameters the manufacturing system as a whole and complex production process chains can be configured. The module is built up in MS Excel™. Thus, it is easily accessible for users who are not familiar with simulation applications. However, basic production configurations for simpler cases can also be done directly in the simulation (which is sufficient for many cases and might be a reasonable option for advanced users).

Parameters

Within the PPC module three different kinds of parameters are considered. Firstly, **general product information** can be set, which play a role in later evaluations for example. These are the (start) weight and a certain value associated with the product. These are for instance the profit margin or the total added value per product.

Secondly, **process module parameters** can be defined through the PPC module (e.g. process time, machine starting time). As mentioned before this is also possible via direct process module configuration whereas for two reasons it might make sense to do it via the PPC MS Excel™ interface. Besides the easier accessibility for non-experienced simulation users this additionally allows configuring product specific process module parameters, e.g. dynamic process

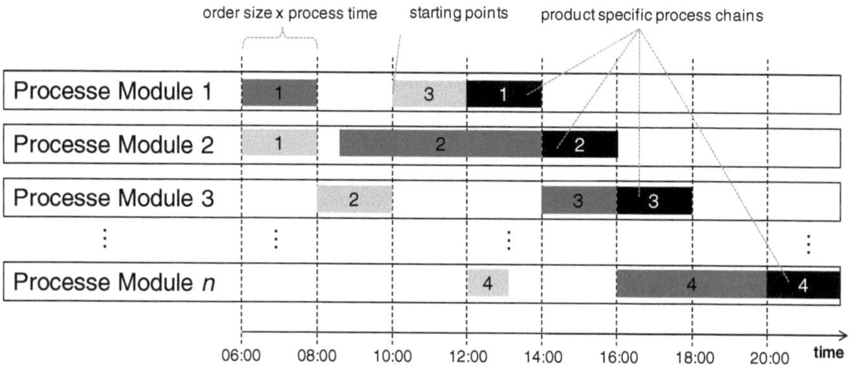

Fig. 69 Input parameters of PPC module

times depending on the actual product. Finally, the **production program** with product specific process chains can be configured as shown in Figure 69.

Besides the aforementioned individual process times, the production program is determined by order sizes and starting points for specific products as depicted in Figure 69. As background for that individual, product specific production process chains can be defined.

Verification

As depicted above, the PPC module works very differently compared to process and TBS modules. It has a superior function and is strongly interacting with both process and MS modules. A separate verification is not possible because this automatically involves all other modules. Therefore verification will be done for all modules together in context of the MS module.

5.3.6 Evaluation/Visualisation (EV) Module

Figure 70 shows the main elements and interfaces of the evaluation and visualisation module. It plays a crucial role for the whole approach as this module brings together data from different other modules.

As inputs, the EV module gets the energy demand of TBS modules and manufacturing system (MS module) as well as further information e.g. regarding production performance. Based on diverse parameters, relevant key figures and diagrams are generated, which are depicting the state of the simulation model as a whole. The EV module is also responsible for transferring data to external tools for the visualisation of energy flows and extended (manual) evaluation.

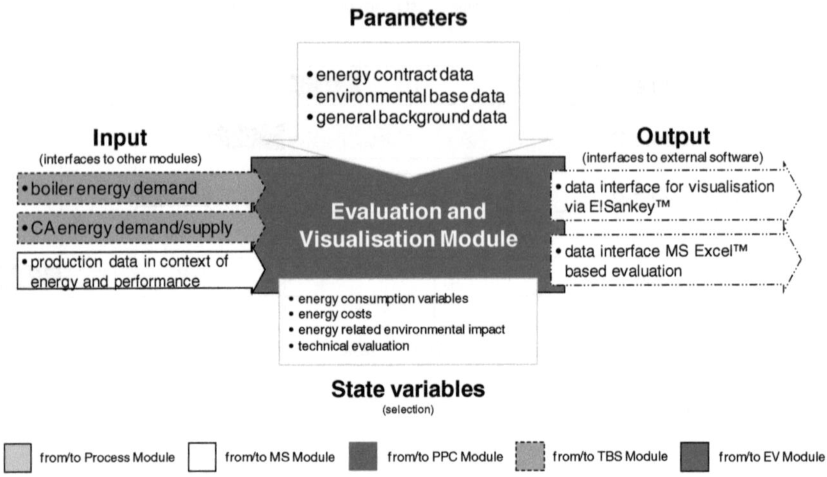

Fig. 70 Inputs, Outputs and Parameters of the EV Module

Parameter

All parameters of the EV module are shown in Table 21. These parameters are mainly background data for the calculation of certain key figures. Therefore no further explanation will be given at this point but later, in context of the calculation of state variables.

State Variables

As mentioned before, the calculation and visualisation of several state variables is the core of the EV module since they depict the state of the whole system. Against the objective as a flexible simulation approach for manufacturing systems the user shall not be limited beforehand. A diversity of variables for evaluation is being offered, which allows the selection of the most appropriate ones for the specific case. Different perspectives are being pursued in this context. Besides technical evaluation and pure energetic view, this includes the calculation of energy costs and related environmental impact. Figure 71 shows a screenshot of the graphical user interface of the simulation environment, which is mainly implemented through the EV module. The relevant key figures will be explained in the following.

Technical Analysis

The evaluation of the general technical capability of a manufacturing system is the usual application for material flow simulation studies. Since the proposed approach aims at an integrated evaluation these issues need to be considered as well. The most important state variables in this context are the product output of the whole system (**production output** - number of finished parts) and the relation

Table 21 Parameter lists of EV module

Parameter	Description	Dimension
CO2_in_kg_per_kWh_electricity	CO_2 conversion factor for electricity usage.	[kg $CO_{2,\,eq}$/kWh]
CO2_in_kg_per_kWh_gas	CO_2 conversion factor for gas combustion.	[kg $CO_{2,\,eq}$/kWh]
CO2_in_kg_per_kWh_oil	CO_2 conversion factor for oil combustion.	[kg $CO_{2,\,eq}$/kWh]
base_electricity_price_per_kWh	Electricity price for base time.	[€/kWh]
peak_electricity_price_per_kWh	Electricity price for peak time.	[€/kWh]
contract_load	Electrical power as defined in contract.	[kW]
dynamic_fees	Various fees depending on consumption.	[€/kWh]
peak_surcharge	Surcharge for peaks over contract load.	[€/kW]
static_fees	Fixed monthly fees.	[€]
daily_worktime	Daily working hours as base for forecast.	[h]
monthly_working_days	Working days as base for forecast	[days]
oil_price_per_kg	Price per kg for oil.	[€/kg]
gas_price_per_kWh	Price per kWh for gas.	[€/kWh]
heating_value_gas	Heating value for oil.	[kWh/m^3]
heating_value_oil	Heating value for gas.	[kWh/kg]
compressed_air_efficiency	Efficiency compressor.	[%]
gas_usage	Usage of gas for steam generation.	1: yes / 0: no
oil_usage	Usage of oil for steam generation.	1: yes / 0: no
electricity_base_load	Base load e.g. sum of small consumers.	[kW]

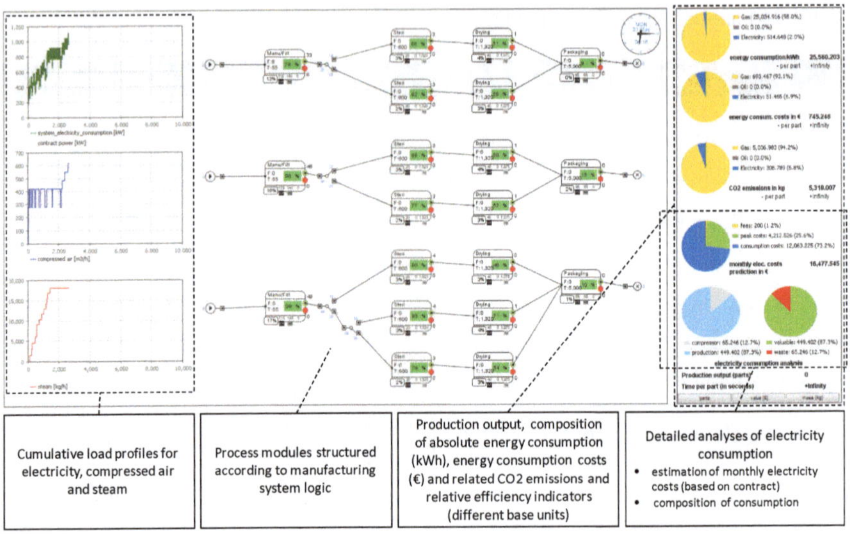

Fig. 71 Screenshot of simulation environment with sample model and diagrams/key figures for evaluation

to time as **time per part**. Additionally, the specific weight and value per product can be defined and therewith production output can also be measured as **cumulated value or weight of finished products**. These are numbers for the system as a whole; on individual process base, further key figures are of importance. Besides the number of **failures** and **waste parts** the **utilisation rate** value plays a key role and was already explained above in context of the process module.

Energy Consumption Analysis

In the context of energy consumption different means for analysis are provided.

The current demand of the whole system on electricity, compressed air and steam are visualised through time based diagrams (**load profiles**) and allow the analysis of consumption patterns over time. The sampling rate is usually one second. For electricity in addition a 15 minutes average is tracked since this reflects the sampling rate relevant for billing purposes.

The calculation of the **total energy consumption** gives an overview from an energetic perspective. Thereby, the focus lies on all forms of energy externally acquired by the company and so it is distinguished between electricity, gas and oil. For visualisation a pie chart is provided, which shows the contribution of all considered forms of energy on total consumption.

Electricity usually plays a major role in context of energy consumption. Therefore further means for analysis are provided. One chart is provided, which shows the shares of **production versus compressor electricity consumption**. Another pie chart introduces the perspective on **valuable and waste electricity consumption**. Electricity consumption is 'valuable' if used for the value creating process. All consumption connected to other operation modes (e.g. idle) is considered as 'waste'. This differentiation helps to identify potentials for improvement. As already explained for each individual process module its current energy demand and the specific contribution to total electricity consumption is also displayed which are strong means for decision support.

Energy Cost Analysis

As Chapter 2.3.1 revealed the pure energetic view is only one necessary but not sufficient perspective. Even more important for a company is the conversion to related energy costs and environmental impact. Energy costs are calculated with two schemes.

Firstly, the costs of energy consumption of electricity, gas and oil are calculated through multiplying consumption with certain price rates which can be determined as parameters. The composition of **total energy consumption costs** is illustrated as a pie chart.

However, Chapter 2.4.1 clearly revealed that at least for electricity a pure conversion of consumption to costs is not sufficient. Electricity costs consist of diverse components depending on different variables. Therefore, electricity supply contract models can be applied in the proposed approach.

- Different consumption price rates for base (6am-8pm) and peak (8pm-6am) times can be inserted in order to reflect daily price changes.
- Typically a negotiated power load limit is part of the contract and can be keyed in. Exceeding this limit results in peak surcharges with a specific rate per kW. Thereby the highest peak per month is relevant.
- Fees can be distinguished in fixed monthly fees and dynamic fees which are depending on consumption. Both types of fees can be considered simultaneously.

Based on this contract data and the current electricity load profile (aggregated to 15 minute interval as usual for billing) the **monthly electricity costs** are being estimated and displayed in a pie chart. While simulation runs are typically covering rather hours or a few days, a similar consumption pattern is assumed for the whole month and forecasted with given parameters in terms of working days and daily work time. The monthly perspective is relevant since this is the usual billing period and certain cost aspects are calculated on this base as well. Altogether, monthly electricity costs describe what companies actually have to pay for electricity and therefore this view is very relevant for consideration.

Energy Related Environmental Impact

For assessing the energy related environmental impact, energy consumption and usage is converted into related green house gas emissions. Typically, **CO_2 or CO_2 equivalents** (which subsumes different GHG) are applied as unit in this context. Necessary conversion factors need to be inserted as parameters. For electricity this factor depends on the specific primary energy mix for electricity generation (see Chapter 2.2). Gas or oil is burned directly at the company in order to e.g. generate heat; in these cases a certain factor for GHG through combustion needs to be keyed in.

Efficiency Key Figures

Besides absolute values the calculation of relative values is a crucial option for multi-criteria evaluation and solving of conflicts of goals. In this context, the usage of appropriate **efficiency figures** is proposed. However, Chapter 2.5 already pointed out that there are difficulties of defining the "right" efficiency indicators for the specific case. The proposed approach does not want to restrict any possibilities for the user, therefore different options were integrated. Either the number of produced parts, the weight of all parts or the economic value of the parts can be used as base. These values are set in relation to pure energy consumption but also energy consumption costs and GHG emissions. Altogether, in accordance with the efficiency definitions given in Chapter 2.5 the following options can be considered:

$$Energy\ efficiency = \frac{production\ output\ (parts,\ weight\ or\ economic\ value}{total\ energy\ input\ (electricity, gas, oil)} \tag{28}$$

$$Energy\ cost\ efficiency = \frac{production\ output\ (parts,\ weight\ or\ economic\ value)}{total\ energy\ consumption\ costs\ (electricity, gas, oil)} \tag{29}$$

$$Energy\ GHG\ efficiency = \frac{production\ output\ (parts,\ weight\ or\ economic\ value}{total\ energy\ related\ GHG\ emissions\ (electricity, gas, oil)} \tag{30}$$

All these efficiency key figures are calculated continuously during runtime. The displayed reference unit (number of parts, weight or economic value) can be switched simply through the graphical user interface.

Evaluation under Uncertainty

Within the proposed simulation approach the consideration of both deterministic and probabilistic models (see Chapter 3.4) is possible. Stochastic effects become relevant if e.g. failure behaviour of machines or certain process times are modelled through distribution functions. As a result, each single simulation run leads to only one possible result in probabilistic models; another run may lead to totally different outcomes. This uncertainty is not acceptable for providing decision support in manufacturing companies. This is specifically true when it comes to comparing different scenarios. It has to be ensured that different outcomes of scenarios are *statistically significant* and not just a result of stochastic effects. To overcome this obstacle and gain acceptable statistical power, a sufficient amount of simulation runs has to be conducted. Variables to estimate the necessary number of runs are (Soper, 2011)

- the statistical power which should be typically >0.80,
- the alpha level or error rate/p-value, whereas a value <0.05 expresses statistical significance

- the effect size d (also named Cohen's d) which expresses the strength of the relationship of two data sets. It is calculated with the mean of the data sets $x_{1,2}$, their variance $s_{1,2}$ and the sample size $n_{1,2}$

$$d = \frac{\bar{x}_1 - \bar{x}_2}{s} \tag{31}$$

$$s = \sqrt{\frac{(n_1 - 1)s_1^2 + (n_2 - 1)s_2^2}{n_1 + n_2}} \tag{32}$$

  For the effect size d three different categories can be distinguished. A value of d=0.2 refers to a small, d=0.5 to a medium and d>=0.8 to a large effect (Cohen, 2009).

Figure 72 shows the necessary number of runs to gain a specific statistical power with respect to the effect size d and an error rate of 0.05 and 0.01 respectively.

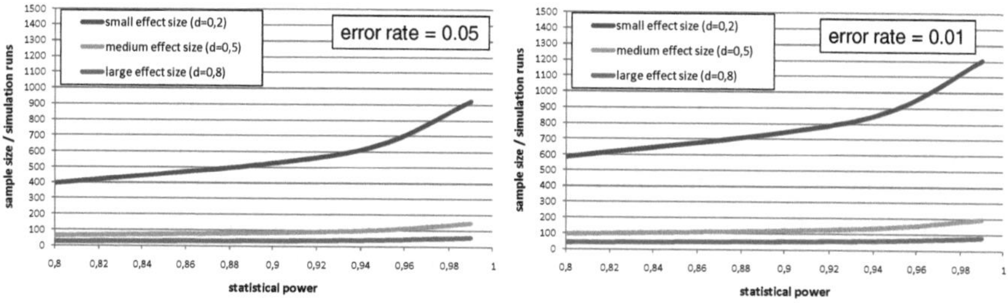

Fig. 72 Necessary sample size depending on effect size, statistical power and error rate (calculated according to Soper, 2011)

The diagrams underline the strong coherences of intended statistical power with sample size (referring to number of simulation runs), effect size and accepted error. Specifically the effect size has a strong influence: this practically means the lower the effect, the more runs have to be conducted to statistically prove the difference between two data sets. The diagrams help to estimate the necessary sample size. For medium and large effects (which are of main interest in analyses) and an acceptable statistical power of 0.8-0.9 with an error rate of 0.05, approx. 20-90 simulation runs are often sufficient to gain reliable results. However, for detailed considerations 500 to >1000 runs might be necessary, which lead to significant effort in terms of computing time. However, any of those analyses are possible with the proposed approach support. The applied software tool AnyLogic offers proper functionalities to efficiently run hundreds of runs (which may take a couple of hours). To ease application, an automatic continuous export of relevant result variables into a MS Excel™ database was integrated, which allows automated simulation and data recording. With this data statistical treatment can be done. Besides the calculation of the statistical confidence, typical key figures for evaluation are the average and standard deviation of the data set. Also the statistical probability to stay below certain limits might be relevant to consider (e.g. monthly energy costs will be <10000€ with a certainty of 95%, see also Figure 73) (Black, 2008; Anderson, 2002).

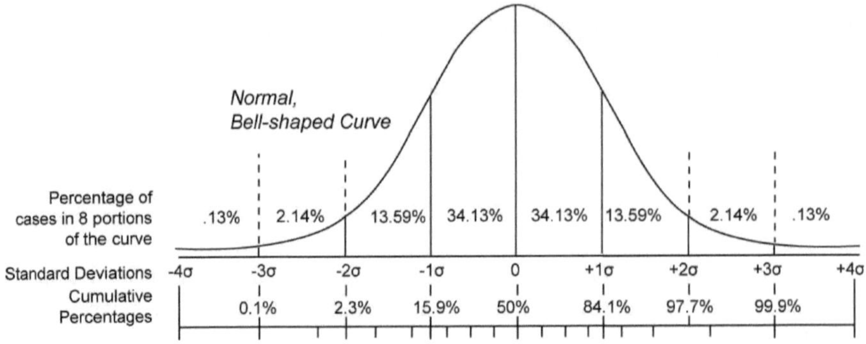

Fig. 73 Selected statistical key figures for a normal distribution (e.g. Black, 2008; Anderson, 2002)

Visualisation of Energy Flows

Besides key figures and charts for supporting evaluation and decision support, general energy flows within the factory can be visualised with Sankey diagrams. Sankey diagrams were developed by H.R. Sankey and are an established method specifically for those kinds of tasks. They can strongly increase the transparency of energy and material flows for complex systems. Therewith, it supports the systematic derivation of most promising improvement measures.

An example for a Sankey diagram was already shown with Figure 14. Another example gives Figure 74 with an old Sankey illustration of a steam plant (which is also the first Sankey diagram) by H.R. Sankey himself (Sankey, 1898).

To understand Sankey diagrams some basic assumptions are important to know (Schmidt, 2006; Schmidt, 2008b):

- variables are related to a specific period of time (in this case simulation time)
- considered values are extensive (different values can be added)
- the arrow width is proportional to its value
- no stocks, what goes in a process has to go out as well

To realise automated Sankey visualisation of simulation results a dynamic interface with the software tool E!Sankey™ (developed by IFU Hamburg GmbH, www.ifu.com) was established. E!Sankey™ is a specific tool for easy visualisation of Sankey diagrams, which are often difficult to draw manually. Every 10 seconds the simulation tool writes energy flow related values in a database, which is being read out by E!Sankey™ and then automatically updates the pre-defined energy flow model.

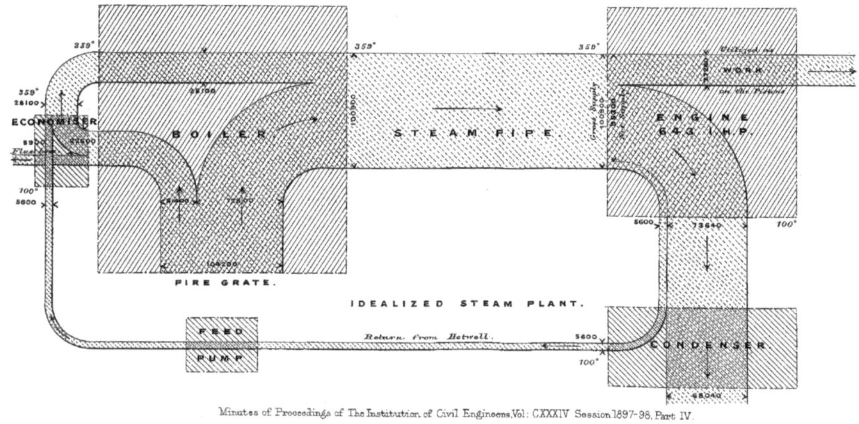

Fig. 74 Example for Sankey diagram for the case of a steam plant (Sankey, 1898 also shown in Schmidt, 2008a)

Verification

The EV module is highly relevant for the system as a whole and strongly interacting with both PPC and MS module. It has a supporting function and its calculations depend on input values of other modules; a separate verification with blending out other modules is not reasonable and hardly possible. Therefore verification will be done for all modules together. If this can prove that calculations work correctly, this module can be considered as verified.

5.3.7 Main Level – MS Module

As depicted in Figure 75 and also earlier in Figure 51, the MS module is the backbone of the whole simulation approach. On this layer all generic modules are logically put together (if not given already when using generic approach configured by PPC module) and parameterised in order to establish an individual model for the specific case. For simulation, the MS module combines process module and PPC data and executes the experiment runs. Thereby, relevant data about energy demand and supply as well as the technical performance of the manufacturing system are transferred to EV, TBS and also back to process modules.

While being highly relevant from a technical perspective as hub for all other modules, the MS module is only a necessary platform and has limited meaning from a content oriented perspective. Several variables are tracked as raw data but directly transferred to the modules where further processing takes place. This simplicity of the MS module is established on purpose: if not accessed via the Excel interface anyway, this layer is the main interface for the user when modelling. To faciliate usage but also for avoiding accidental changes of critical model code of this module, most functionalities were inserted in separate modules and the MS module itself was kept simple and transparent.

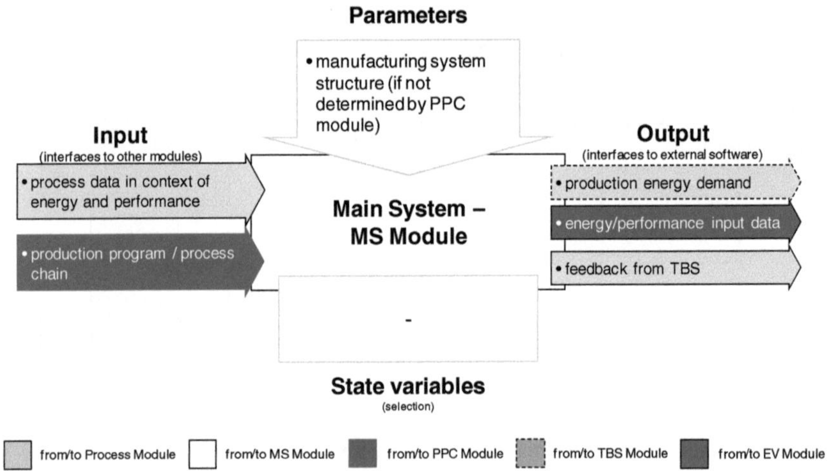

Fig. 75 Inputs, Outputs and Parameters of the MS Module

Verification

As indicated before, due to their close relation and interactions the three modules EV, PPC and MS shall be verified all at once at this point. A small case is built up, which allows checking functionalities and calculation schemes connected with those modules. This case will consist of three process modules and is deterministic in nature, which allows transparent assessment of model behaviour and static calculation of state variable for comparison with simulation results. The manufacturing system is completely configured with the PPC module over MS Excel™ while using the generic simulation approach. There is no rigid linkage between the machines but a free flow of products through the system (just determined by PPC). An order of 10 parts shall be produced on three different machines in defined sequences. Certain process times, a simple energy consumption pattern (idle and processing), and starting points were determined for each machine. For reasons of simplicity, only electricity consumption was considered. However, calculation patterns etc for other forms of energy are basically the same. Figure 76 shows the configuration of the verification experiment and the results from static calculation as well as from simulation.

The simulated electricity load profile reflects the activities which were planned in the PPC module. All processes start and end as intended (delays through e.g. failures were not considered). Basic machine data like state based power demand or process time was transferred correctly. Therewith the PPC module can be considered as verified. The comparison of static calculation and simulation outcome reveals that both results in similar values for energy consumption and production output. These are the main variables as all other variables are based on the correct calculation of those figures. With these positive results, the EV module can also be considered as verified. The MS module is the hub and basically the

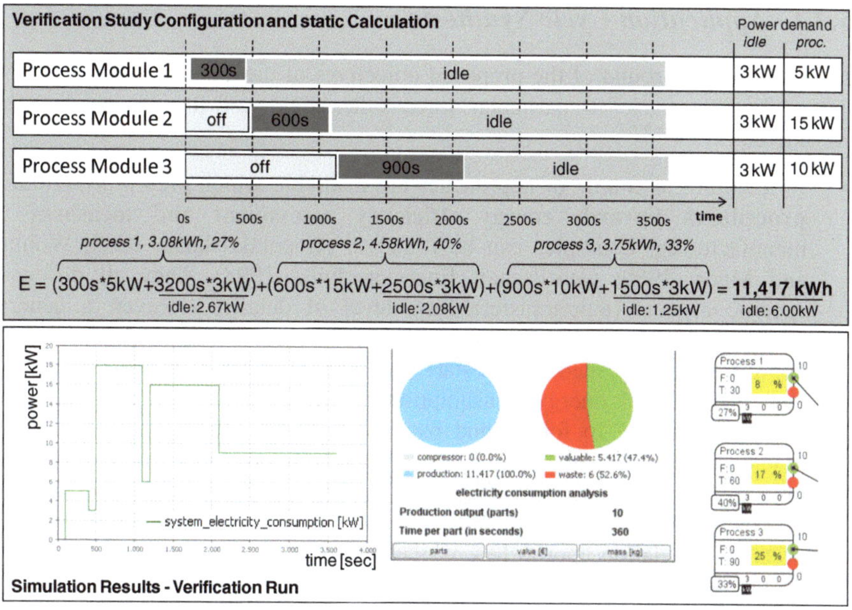

Fig. 76 Verification results for MS, EV and PPC module

enabler of all those functionalities. Hence, due to the positive verification of all other modules, the MS module is also verified. Additionally, the results proved again the correctness of the process modules and their interaction with the MS module.

5.4 Application Cycle

As an integral part of the concept as a whole an application cycle will be developed. It shall

- support users to independently (without external help) and **systematically identify potentials to improve the energy efficiency** in their specific manufacturing company.
- guide users through the **goal-oriented and correct application of the simulation** approach for their specific case
- enable users to **derive and evaluate different measures** with the proposed simulation approach

The development of the application cycle and the description of its individual phases will be presented in the following.

5.4.1 Application Cycle Synthesis

Against the background of the proposed objectives of the concept as a whole and the application cycle in particular three other fields of interest need to be considered:

- In literature, a couple of approaches are available which present **systematic procedures towards energy efficiency assessment and measures** in manufacturing. Examples can be found in (Gopalakrishnan, 2005; Wohinz and Moor, 1989; Gesellschaft Energietechnik, 1998). They all differ in specific aspects (e.g. transferability, level of detail); however, a general structure can be identified, which is shown in Figure 77. Typically top-down-approaches are pursued, which start with a rough analysis of existing energy data and general energy consumption of the company. Afterwards more detailed steps are to follow and based on estimated energy (cost) savings potential measures are prioritised and implemented. Due to their similar objective these procedures can certainly give a good orientation for the proposed application cycle. However, none of them incorporates energy oriented simulation approaches, which underlines the necessity for a further development.
- In recent years the development and implementation of **energy management systems** has become more and more the focus of companies. "An Energy Management System (EnMS) systematically records the energy flux and [...] helps a company to continuously and systematically improve its energy performance while taking into consideration other relevant and legal requirements. An EnMS includes the organisational and informational structures required for implementing energy management, including resources" (Federal Ministry for the Environment, 2010). Based on previous experiences from different countries (e.g. VDI, 2006) with DIN EN 16001 a European standard was introduced in 2009 (DIN EN 16001). On this base, an international standard is planned for 2011 with ISO 50001 (Pinero, 2009). DIN EN 16001 reflects the Plan-Do-Check-Act (PDCA) cycle for continuous improvement. Its main phases are displayed in Figure 77. In contrast to the systematic improvement procedure presented before it is broader in nature and explicitly includes extended aspects like energy policy development, organisational and infrastructure setup, auditing/reporting or communication. The proposed application cycle does not aim to include these issues; it should rather be compatible with EnMS initiatives and support their implementation and operation.
- In the proposed application cycle the **necessary steps for conducting simulation studies** need to be respected (Banks, 2010; Law, 2007; Wenzel et al., 2008; VDI, 2007). A procedure was already introduced in Chapter 3.4 and is display again in Figure 77. Main steps are the definition of the system/problem, data acquisition and modelling, verification and validation as well as the iterative conduction of experiments towards the implementation of promising measures.

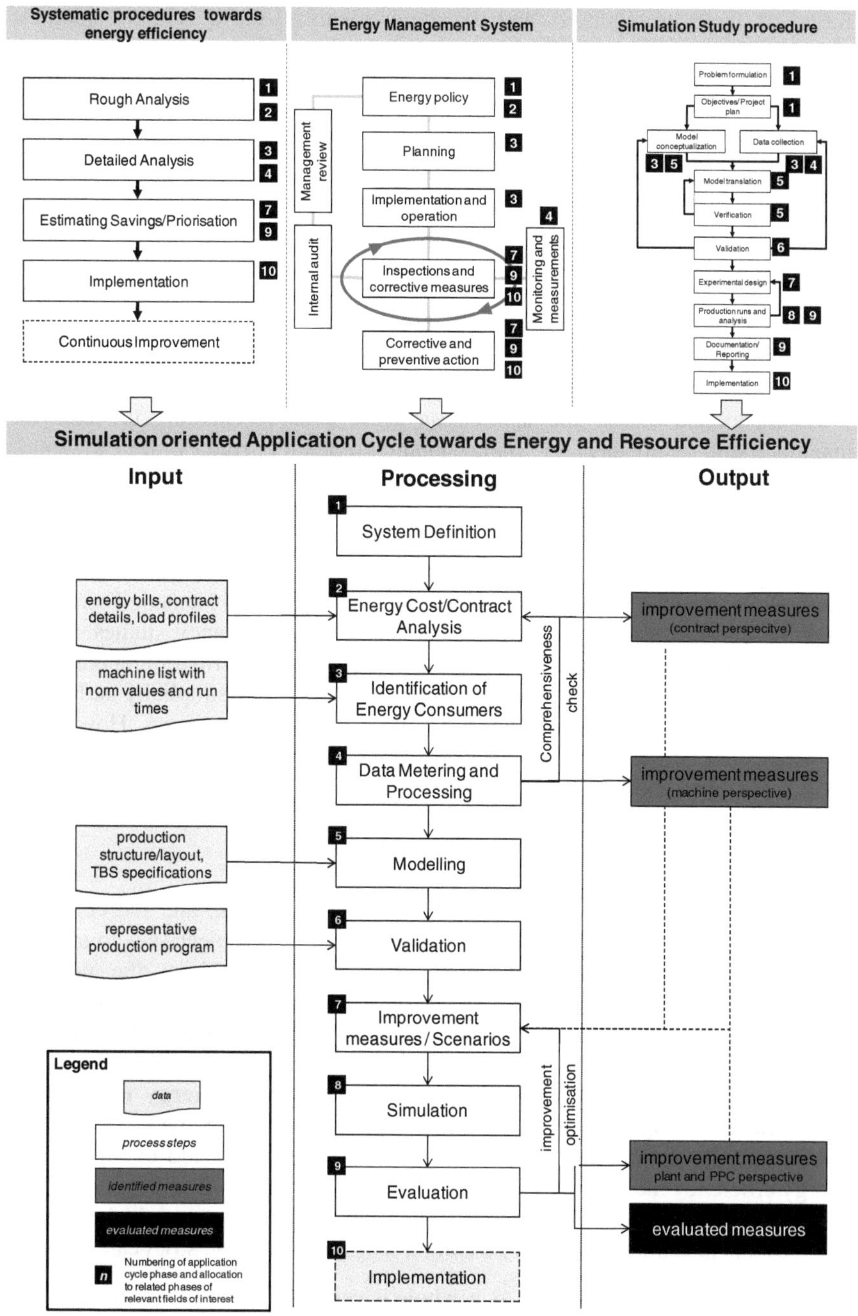

Fig. 77 Synthesis of proposed application cycle

Based on these prerequisites and the requirements of the developed simulation approach the proposed application cycle is shown in Figure 77. The figure also shows the allocation and contribution of certain phases to the mentioned fields of interest. As depicted, the application cycle brings together all necessary actions towards a simulation based improvement of energy efficiency.

In general, it is also a top-down procedure which starts with a general analysis of energy costs and flows. In the next step energy consumption as a whole is broken down to single consumers whose consumption patterns need to be measured in detail. This data serves as input for modelling the considered system with the developed simulation approach. Since the simulation environment itself is already verified, only the validation for the specific case is necessary in the next step. With the validated model simulation runs can be conducted, which allow deriving and evaluating diverse measures for energy efficiency improvement.

As mentioned above, the application cycle aims at enabling the user in a manufacturing company to independently conduct studies for systematic derivation and evaluation of energy efficiency measures for his specific case. Thereby the usage of the developed energy oriented simulation approach is motivated. The user is guided through the whole application cycle. No specific requirements are connected to the first implementation. In the following each phase and connected tools of the application cycle will be presented. Thereby it is assumed that the user has no experiences in energy efficiency studies. Besides some standard documents (e.g. energy bills) no further information about energy consumption is needed at the starting point. For reasons of simplicity and general relevance the explanations will have a certain focus on electricity. However, as also indicated before, all forms of energy can be considered with the presented application cycle and its inherent tools.

5.4.2 Step 1: Objective and System Definition

The very first step is the definition of the designated task, and the considered system as well as connected system boundaries. The usual objective is to minimise energy consumption while still being able to fulfil at least the necessary production tasks. This leads to the objective of **increasing energy efficiency** with the aforementioned definition as ratio of production output to energy input. Companies are certainly less interested in improving these issues from a pure energetic point of view – typically the **minimisation of energy costs and/or environmental impact** with given output are the main objectives. This may be connected with defined quantitative targets for a specific period of time, e.g. an energy efficiency increase of 5% in one year.

In context of system definition, the following alternatives might be possible whereas this is neither necessarily a complete list nor are these strict categories:

- The analysis of a **company as a whole** is certainly an ideal case because total energy costs and all energy carriers are considered. Therewith the danger of problem shifting and focusing on minor relevant issues is at least minimised, potential cost reduction effect is not restricted and might be optimal and there is a maximum leverage/synergy effect over the whole company (e.g. measures can be applied in different areas of the company). However, the complexity of analysis, connected efforts in terms of costs and time as well as the uncertainty of considerations might increase strongly – specifically for large companies with several complex production structures.

- Therefore, concentration on a certain **area of production** (e.g. specific factory/manufacturing system) can be a promising approach. This might compensate some of the challenges mentioned above which might restrict some opportunities.
- Considering a **process chain for a specific product** can provide an alternating perspective for the analysis. This might involve several processes in one or even over diverse companies. Typical questions would be to minimise the environmental impact (e.g. carbon footprint) or the associated costs of product. Whereas the calculation of direct consumption is not a problem, the allocation of indirect consumption has to be addressed carefully while it may also be related to different products.

Diverse variables may influence the system definition. The main drivers are the energy related objectives of the company (e.g. energy cost decrease, improving the environmental impact of a specific product) but also data availability or technical as well as organisational structures (e.g. size of company, specific structure of cost allocation, different production sites of a company) can influence this decision.

5.4.3 Step 2: Total Energy Consumption and Contract Analysis

In the next step the energy costs and the energy related environmental impact of the selected system need to be assessed. This includes all kinds of energy acquired from external sources – for most manufacturing companies, electricity plus gas and/or oil. Base for these analyses are the related bills from energy suppliers (which might be internal suppliers when being different cost/profit centres) as well as their underlying contracts. Whereas the costs for gas and oil are quite clearly defined depending on consumption, the electricity cost calculation can be more complex as already shown in Chapter 2.4.1. Hence, the electricity supply contract should be analysed carefully and the composition of electricity costs has to be determined. For the determination of energy related environmental impact conversion factors for electricity usage and combustion of gas or oil can be used. Altogether the following questions can serve as guideline for this step:

- Which forms of energy are relevant for the specific case?
- What is the contribution to total energy consumption (in kWh), costs (in e.g. €) and environmental impact (in e.g. kg $CO_{2,eq}$) for each energy form?
- Which cost portions contribute (and to what extend) to total electricity costs?

This information gives clear indications for further priorities; depending on the outcomes next steps shall concentrate on the most important forms of energy. Besides the necessary prioritisation for the next steps, from an economic point of view first fields of action can be identified through contract analysis. Figure 78 gives an overview of means to influence electricity costs for the case of Germany.

electricity cost driver (different cost portions)	energy acquisition		energy usage	
	influence through pricing/contract	no influence	influence through usage pattern	no influence
Taxes		x	x	
Licence fee		x	x	
Renewable Energy Sources Act		x	x	
Combined heat and power generation Act		x	x	
Measuring price		x		x
System usage fee/network access	x		x	
Price/costs for electrical work	x		x	
Price/costs for electrical power	x		x	
Base/peak price; mixed price	x		x	
Reactive energy price	x		x	
Minimum/contract power	x			x
Maximum load utilisation period		x	x	
Load factor		x	x	

Fig. 78 Matrix for means to influence electricity costs

As mentioned before, electricity costs consist of manifold components, which may be influenced through modifying contract conditions through negotiating with the supplier or influencing consumption behaviour in the company (which is addressed in later steps of the application cycle). Since there are almost no standard contracts in industry, specifically for large consumers there are opportunities to consciously influence contract contents. Typical examples are the determination of a mixed price rate (instead of having different rates for peak and base times), contract load or the ratio of consumption related price rate to peak demand surcharges. Due to the complexity of the problem the evaluation of alternatives is difficult for the user. However, basic possible influencing variables have to be identified at this stage while alternatives can be evaluated with the simulation approach at a later stage.

Additionally, time based analyses of energy consumption have to be conducted. Therefore monthly consumption values (from energy bills) can be used to gain an overview of e.g. seasonal influences. Even more, for electricity it is possible to get the 15 minute load profiles for specific periods of time. This gives a good hint regarding the electricity consumption pattern of the system and may reveal first saving potentials (e.g. electricity consumption at night or weekends).

As an example, Figure 79 shows a sample electricity load profile for two months of a sample manufacturing company (SME). In combination with the yearly consumption and the details of the contract model some interesting aspects can already be observed. There are very few distinctive energy peaks which exceed 200 kW – evidently in the morning hours when machines are turned on simultaneously. A reduction of these few peaks (which might be easily avoidable if being aware of the impact) would lead to cost saving of approx. 600€ per year. Additionally, it can be observed that the base load is at least approx. 5-6 kW. Assuming that this load is demanded 24 hours a day and 365 days a year it gets clear that this alone accounts for approx. 15% of total electricity consumption without any value creation.

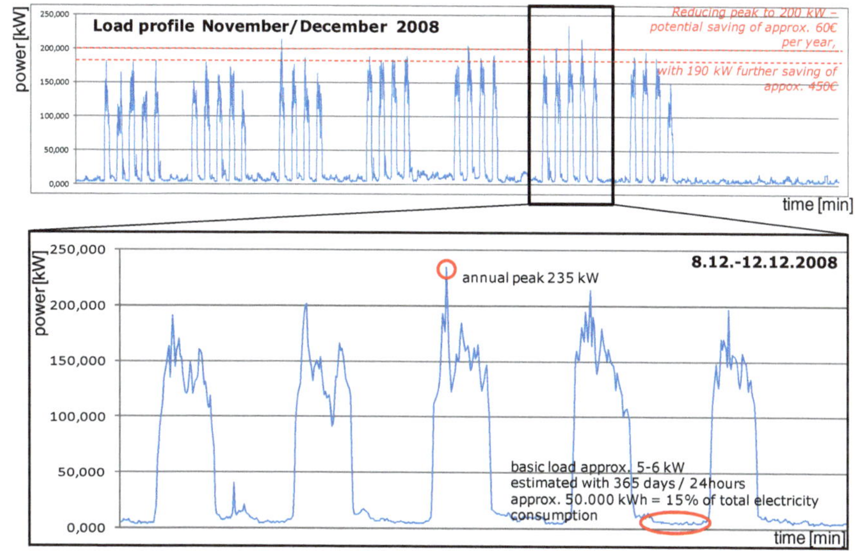

Fig. 79 Example load profile of manufacturing company

All these considerations are most useful when considering the company as a whole since all consumers together and the specific contract model determine the energy costs. However, those analyses might also give interesting hints if only specific areas of production are considered. On the one hand, realistic energy costs calculation schemes (e.g. price rates) can be assumed. On the other hand, later on, the contribution of the considered area to total energy consumption can be estimated.

5.4.4 Step 3: Identification of Energy Consumers

While now knowing the important energy flows of the system in general, they have to be broken down to single consumers in the next step. As it is assumed that no detailed consumption information is available, an alternative way for identification and first prioritisation is needed. For this purpose, a list of all technical equipment of the considered system needs to be generated. This involves production machines but explicitly also includes supporting machines like compressors. Besides general data (like type, name, year of construction etc), two values are critical to gather in this context:

- The **nominal value** for energy consumption (nominal energy demand) of a machine can be found in the technical documentation and/or on type plates directly on the machine. As mentioned before, specifically in the case of electricity the nominal value tends to be too high and rarely reflects the real power demand in operation. However, it can serve as base for first estimation.
- The **operation time** per year is a usual machine related key figure and available in ERP systems or manual calculations of e.g. utilisation rates.

With those values an estimated energy consumption value can be calculated for each machine of the system.

$$\begin{aligned} \textit{yearly energy consumption}_{estimated} = \textit{nominal energy demand} * \\ \textit{operation time per year} \end{aligned} \quad (33)$$

Therewith the machines of the system can be sorted according to this value which is giving a clear indication of priorities for the next steps. It is crucial to ensure that all potential consumers of the system were considered. An example for the case of electricity is shown in Figure 80 (based on a real company case). Evidently, electricity consumption is dominated by few consumers: in this case just six machines sum up for half of the estimated consumption, approx. 15 machines account for 80%. With not even 30 machines determining consumption, the majority of consumers are most likely not worth being considered in detail (diagram is even cut off; altogether 78 machines are in the list).

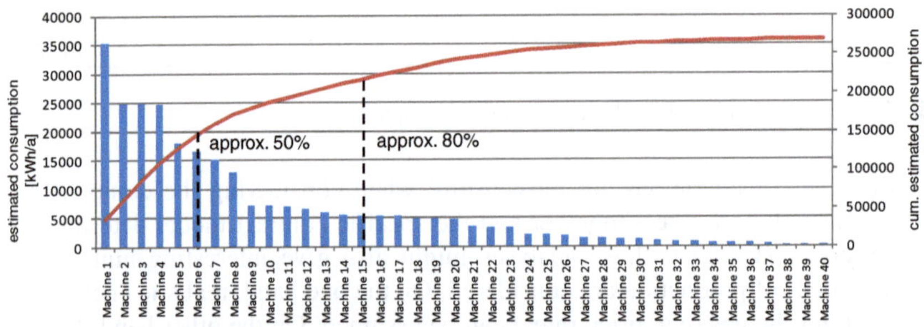

Typ	manufacturer	operation time	nominal value	estimated consumption
		h/a	kW	kWh/a
Machine 1	XY	736	48	35328
Machine 2	ABC	2070	12	24840
Machine 3	CY	920	27	24840
Machine 4	XZ	800	31	24800
Machine 5	XY	8200	2,2	18040
Machine 6	Z	368	45	16560
Machine *n*	...	...	...	...

Fig. 80 Example for estimation of electricity consumption with pareto analysis

For further considerations and distinctive decision support this data is transferred into an energy portfolio as shown in Figure 81 in general and also as an example. Therewith the composition of consumption in its determining parts power demand and operating time is presented separately. This allows the classification of energy consumers into four distinctive categories, leading to different strategies for further actions:

- High power demand, low operating time (category I): these consumers may be critical if peak surcharges make up a significant amount on energy costs. Detailed measurements should be made to understand if and in which operation mode demand peaks occur. Means for improvement can be technically or organisational in nature, examples are an energy sensitive production planning and control or load management.
- High power demand, high operating time (category II): these consumers are most critical as they are responsible for most of the energy consumption and should be analysed in detail through certain measurements. Means for improvement will typically be more technical driven, e.g. improvement of machine components to reduce energy demand.
- Low power demand, high operating time (category III): these consumers can contribute significantly to total energy consumption due to their relatively long operating time over the year while demanding little power (a pump or electric driver on conveyor). Detailed measurements are often not even necessary. These kinds of consumers might offer some feasible approaches for improvement: with relatively low necessary investment and high leverage effect through operating time, appropriate amortisation times are very likely.
- Low power demand, low operating time (category IV): these consumers are less relevant in context of the energy consumption of the system. No further efforts should be spent for detailed consideration.

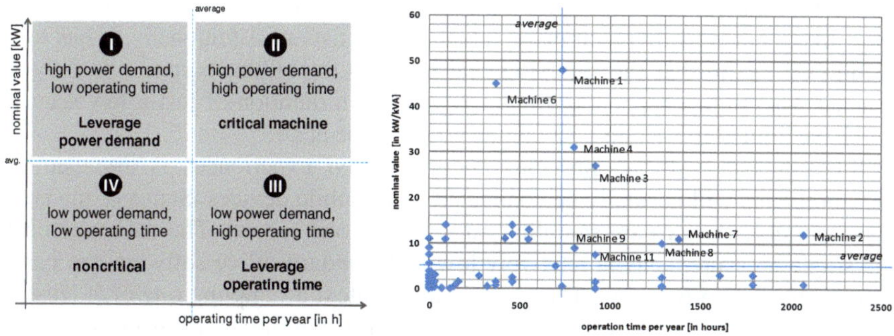

Fig. 81 Energy portfolio as a tool for classifying energy consumers

5.4.5 Step 4: Data Metering and Processing

Previous classifications already gave a clear indication about priorities for detailed energy consumption measurements. Typically, consumers of categories I and II should be considered in further detail. In the next step the necessary accuracy for this measurement needs to be determined. Against this background, Figure 82 shows the influence of different sampling rates in context of the accuracy of resulting energy consumption data. It reveals that too rough measurements might

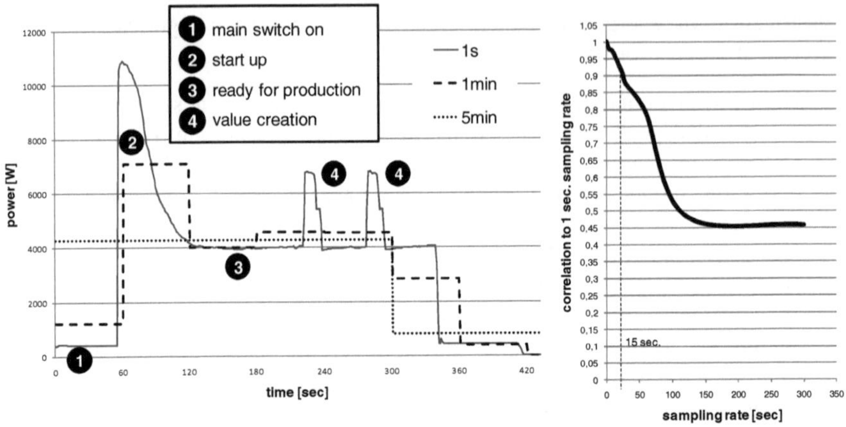

Fig. 82 Influence of different sampling rates on accuracy of energy consumption patterns

lead to a significant loss of information. A later allocation of consumption to different operation states is hardly possible. For the same measurement, the figure also shows a correlation analysis compared to data measured with a sampling rate of one second. It reveals that differences to the measured curve are fast becoming strongly significant. To achieve a correlation value of 0.9, the necessary sampling rate should not be larger than 15 seconds.

A lot of further variables influence the determination of an appropriate measuring strategy. As one example, smaller sampling rates lead to a larger amount of data that has to be handled. If process times are long, extremely detailed measurement will most likely not provide much more valuable information. Additionally, it has to be mentioned again, that the sampling rate relevant for billing purposes is only 15 minutes – so things happening irregularly and with duration of just a few seconds will have limited impact on a 15 minute power value at all.

Since a user with no previous energy related experiences is assumed, all measurements will normally be conducted with mobile devices. States of the art of those portable devices are sampling rates around one second. The data volume of the single measurements is relatively low and standard office software can easily be used for data processing. Therefore in this context a sampling rate of at least 10 seconds (better using 1 second) is suggested (for continuous measurement other criteria might lead to different suggestions).

With measured energy data for all machines, the priority list with estimated yearly energy consumption as well as the energy portfolio can be updated. With this now more realistic data the contribution of considered machines on real total power demand and energy consumption (derived from step 2) can be calculated. This ensures the completeness of the analysis and also serves as criterion for stopping measurements (e.g. if 80% of energy consumption is caused by machines already considered further measurement activities are not necessary).

Besides laying the necessary data background for further steps, the analysis of energy consumption behaviour directly helps to derive machine based measures for improvement. Obvious saving opportunities can often be identified and employed at once. Again it shall be mentioned that this is also true and relevant for other forms of energy: whereas certainly being different in terms of measuring equipment and strategy those analyses are relevant and applicable for e.g. compressed air, steam or gas consumption as well.

5.4.6 Step 5: Modelling

In the next step the considered manufacturing system will be transferred into the proposed simulation environment. Practical and technical aspects of modelling were already presented in detail in Chapter 5.2. At this point, not the question 'how to model?' but rather 'what to model?' is addressed. This directly leads to the determination of the necessary level of detail.

Besides the data being gathered so far, further information on production machines and structure as well as technical building services is necessary.

- For production machines at least process times for the considered products are necessary. Information on failure behaviour (MTTR, MTTF) will be helpful for more realistic considerations. Additionally, product specific process chain sequences are crucial for modelling – it has to be clarified which processes determine the material flow of certain products within the system. Value stream maps (VSM) are an ideal base for modelling because they illustrate both, process times and sequences.
- For the relevant equipment of technical building service further system specifications and operational data are necessary. In the case of compressed air these are e.g. the system volume, nominal air pressure, amount and technical data of compressors and their modes of operation. More details on necessary parameters (also for steam generation) can be found in Chapter 5.2.

As already mentioned in Chapter 5.2, the proposed simulation approach is very flexible in terms of level of detail. While extensive models increase the complexity of analysis and necessary computing time, it is beneficial to concentrate on most relevant machines. Figure 83 shows a decision tree that supports the selection of an appropriate level of detail for modelling each consumer. While keeping in mind the energy related relevance of each consumer (classification in categories), different general alternatives can be distinguished.

An **individual process module** is necessary if the process is important for the material flow as a whole and integration with other consumers is not possible or feasible (e.g. consumer has to be treated separately to analyse behaviour in detail). In many cases it is favourable to combine consumers to **integrated process modules**. This is possible if e.g. processes are directly connected in the process chain (e.g. neighbouring processing) or there is a supplying process whose energy consumption is related with the operations of another process (e.g. machine related cooling device). Practically, energy consumption is just added to the consumption of the main process in this case. A **separate TBS module** should be modelled if the consumer does not have a constant consumption and is responsible for supplying the system as a whole (e.g. compressed air system). As described before it is possible to subsume less relevant energy consumers as constant **base consumption** of the system.

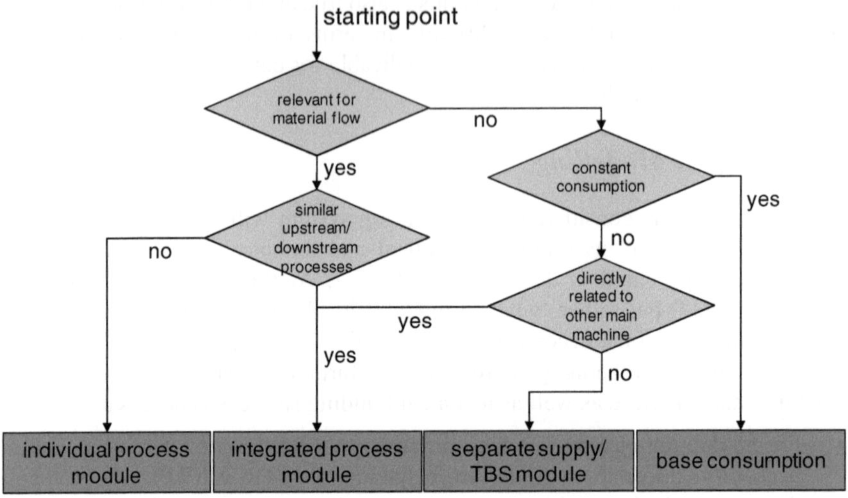

Fig. 83 Decision tree for level of detail while modelling

5.4.7 Step 6: Validation

Verification and validation are essential issues when conducting simulation studies. Verification is ensured already since certain studies were conducted for each single module in the development phase. However, the validity of the simulation model has to be assessed for the specific case. Since validation is the assessment whether "the model behaves with satisfactory accuracy consistent with the study objectives" (Banks, 2010). Several techniques are available (see Figure 35); typically it is realised through comparison of a simulation run with the comparable situation of the real system.

In this context the ideal validation background would be the detailed consumption data of the system (load profile and total energy consumption) for all relevant forms of energy plus the real production data (production program and individual machine behaviour) for the same period of time. This would allow conducting simulation runs with the given production program and comparing both energy as well as performance related target variables for validity check. However, in practical cases, specifically energy related data will be rarely available in the designated form and even less in combination with related production performance data. Against this background also an isolated consideration can be considered as sufficient:

- **Validation with given production data:** with the given production program validation runs can be conducted. Results in terms of selected key figures like output or utilisation are compared to real production data. If system behaviour is clearly determinable (e.g. automated production line with rigid linkage) it is also possible to exclude probabilistic effects to test model behaviour ("Fixed Value Test" - Rabe et al., 2008, e.g. theoretical output rate can be calculated statically and compared to simulation results). If the system behaviour is correct (e.g. similar production output), the model can be considered as validated. Since energy consumption patterns rely on measurements and the simulation environmental itself is verified also energy consumption behaviour of the system can be assumed to be correct.

- **Validation with given energy data:** vice versa - it might be possible that only some energy related data is available, e.g. the energy consumption or a load profile for a certain period of time. However, there might be no related production data. In this case the simulated energy demand and consumption should be compared with given data. Additionally, through careful checks of assumption and process times as well as critical review with experts the production performance related validity should be ensured.

5.4.8 Step 7: Scenario Building

The validated simulation model can now be used to derive and evaluate different improvement measures towards energy efficiency, which leads to different scenarios being considered (the validation run can serve as base scenario for comparison). Thereby, **single production machine and TBS measures** derived earlier can be tested in context of their impact on the system as a whole. It can be assured that no negative consequences are occurring on system level (or, vice versa, positive effects occur) and the contribution towards improvement from a holistic perspective can be determined. This helps to prioritise those measures. Furthermore, there are specific **measures on factory/manufacturing system** level (e.g. load management, energy aware production planning and control). Their functionality and impact on diverse target variables can be tested as well. As described earlier, one of the target criteria is the energy cost estimation based on realistic contract models. Vice versa, this aspect also allows the testing of **alternative energy contract models** against the background of the behaviour of the manufacturing system. Finally, also **optimisation studies** are possible while using the solver OptQuest™ which is integrated in AnyLogic™. Therefore it is necessary to define a target variable (e.g. minimising energy costs), certain frame conditions (e.g. ensuring defined production volume) and parameters (e.g. speed of production) that can be changed.

5.4.9 Step 8: Simulation Runs

In an iterative procedure, diverse simulation runs are conducted, which reflect different scenarios as formerly described. The results of each run should be written down and transferred into e.g. MS Excel™ to be able to conduct extensive evaluation afterwards. If the model contains probabilistic elements a sufficient number of simulation runs need to be conducted. Against the background of the considerations from Chapter 5.3.6 the conduction of 50 simulation runs is a reasonable way to obtain a good estimation of value distribution. Through using the functionalities of AnyLogic™ it is possible to conduct these runs automatically through the function "parameter variation"/replications. This may take several hours - thereby, relevant key figures are automatically written in MS Excel™ as raw data to enable later statistical treatment and evaluation. For instance, with the equations given it is possible to calculate the effect size *d* for two scenarios. Herewith and the intended statistical power (typically at least 0.8) as well as error rate (by convention maximal 0.05), the necessary amount of simulation runs can be determined according to Figure 72. This allows identifying whether the differences of two runs are proved to be statistically significant or further runs are necessary.

5.4.10 Step 9: Evaluation

As presented before diverse key figures and diagrams support transparency and the derivation of promising fields of action. For testing improvement measures, the evaluation and comparison of different scenarios is necessary. Once having identified a promising measure the economic profitability has to be ensured through appropriate calculations.

Evaluation and Comparison of Scenarios

Several alternative target variables are provided which may serve as base for evaluating scenarios. The set of relevant variables is individual for the specific case and subject of e.g. study objectives (step 1). While several targets are addressed simultaneously, multi-criteria evaluation including balancing of conflicting goals might be possible. Several sophisticated methods for Multi-Criteria Decision Making (MCDM) are available in literature (e.g. PROMETHEE by Brans et al., 1986). For reasons of easier applicability in industrial practice a simpler approach is proposed here. Sample results of scenario studies are shown in Figure 84. In this case with energy consumption, energy costs and production performance (in terms of necessary time for 100 parts) three target variables were considered (all of these variable shall be as low as possible). Each variable is associated with a certain weighting factor and normalised to relative values (compared to base scenario).

Therewith, the weighted sum can be calculated and serve as a base for decisions.

$$weighted\ sum = \sum_{target\ categories} weighting\ factor * relative\ value\ of\ target\ category \quad (34)$$

Scenario	Description	Energy consumption (w. factor=1)	relative values	Energy costs (w. factor=1)	relative values	Prod. time (100 parts) (w. factor=1)	relative values	weighted sum	Energy efficiency consumption/parts
Base	Standard scenario	10000 kWh	1.00	1000 €	1.00	200 seconds	1.00	1.000	100 kWh/part
1	Improved machine	9000 kWh	0.90	900 €	0.90	250 seconds	1.25	1.013	90 kWh/part
2	Alternative PPC	7500 kWh	0.75	1200 €	1.20	150 seconds	0.75	0.675	75 kWh/part
n	...	...	...	...	...	...	...		

Fig. 84 Sample evaluation of simulation results

These results can also be depicted via a spider diagram as shown in Figure 84. This supports decisions while showing all targets and scenarios graphically with one view. While all target variables are supposed to be as low as possible in this example, the goal is to minimise the area of the triangle in this case.

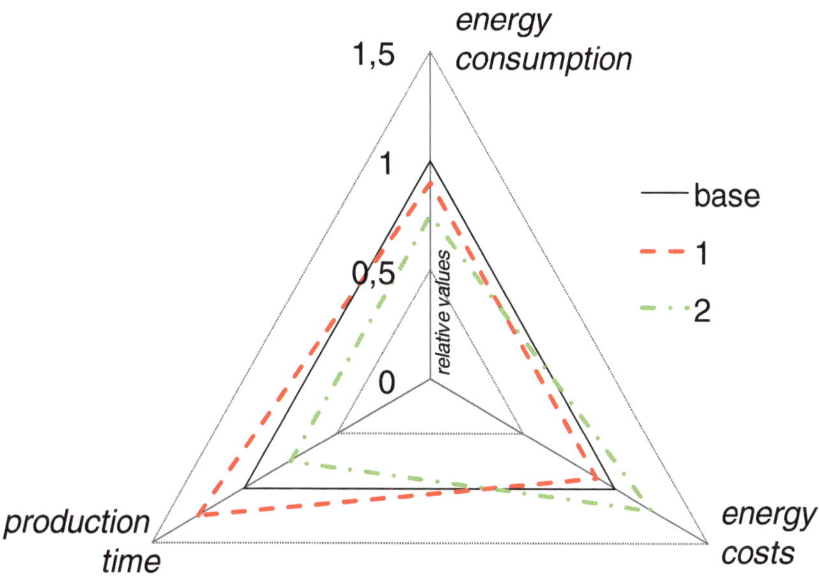

Fig. 85 Graphical representation of simulation results

As the sample results show, conflicts of goals between different target variables might occur. Besides the weighted sum of all target variables, efficiency indicators (see Chapter 5.3.6) can be used to bring together values and balancing targets in order to assess the favourability of scenarios. These key figures are often more intuitive and easier to use but may not be sufficient if manifold target variables have to be considered simultaneously. A hierarchical combination of relative and absolute values might be a promising way. For example, the absolute achievement of most critical goals, like production output can be assured in the first step. Scenarios that pass the defined threshold are compared afterwards with respect to e.g. efficiency indicators in order to identify the best solution.

The consideration shows that a general recommendation is hard to give as it is always a matter of the specific study. However, the presented means for evaluation are generic in nature, which allows a broad applicability for all cases.

Investment Decisions

Measures towards energy efficiency are usually connected with necessary investment. After having identified promising alternatives from a technical and energetic perspective, their profitability has to be calculated before implementation. Diverse calculation schemes and criteria are available and shall not be addressed in detail here. Examples are the calculation of the payback period or the net present value, which includes financial aspects like interest rates (Brigham and Houston, 2009; Khatib, 2003). The simulation results can directly be used in these calculations while e.g. very specific and realistic energy cost (savings) data can be provided.

5.4.11 Step 10: Implementation

The technical and/or organisational implementation of promising measures is certainly the ultimate goal of the proposed concept but rather a practical issue and therefore not explicitly focused. After implementation it has to be ensured that the simulation model is also updated as this serves as the new standard for further simulation studies.

Chapter 6
Application of Concept

In this chapter the developed simulation based concept for improving energy efficiency will be applied to several case studies in order to underline applicability and potentials of the approach. Thereby different cases are being considered to show the opportunities. Each case will follow the steps of the presented application cycle. While the developed concept is generic in nature, the examples will underline case specific implementation and practical implications of each step. The case studies are based on real data from several companies; for reasons of confidentiality many values will just be given in a relative manner.

6.1 Aluminium Die Casting

The first case study focuses on a company which produces die casted parts for the automotive industry (Herrmann et al., 2011a). Products are manufactured in mass production with continuous material flow in transfer lines with rigid linkage (highly automated). Production is running in a three shift system (24 hours a day).

Step 1: Objective and System Definition

In this case study one specific transfer line for the production of a specific product shall be considered in detail. The objective is to minimise direct electricity costs while still obtaining the same output on products.

Step 2: Total Energy Consumption and Contract Analysis

Production machines in the designated area directly consume electricity as well as compressed air. However, for the purpose of this study just direct electricity consumption is considered. The actual electricity consumption of this area is not known due to missing measuring points. The considered transfer line is part of a large production site with diverse facilities. A separate bill is not available. In terms of costs, only consumption induced cost portions need to be considered (no peak surcharges or fees).

Step 3: Identification of Energy Consumers

The production area that is analysed is clearly defined. No further prioritisation is necessary – the objective is to analyse a specific production line as a whole, which

S. Thiede: Energy Efficiency in Manufacturing Systems, SPLCEM, pp. 145–169.
springerlink.com

does not contain too many different consumers. Therefore all consumers will be considered. An overview of the system structure and all machines involved is shown in Figure 86. Three parallel die casting cells (which contain robots for spraying and handling as well as heating/cooling units) convert liquid aluminium into the raw part which is further processed through a saw. Afterwards, parallel CNC milling takes place on two parallel machines before abrasive blasting and palletising finalise the process chain.

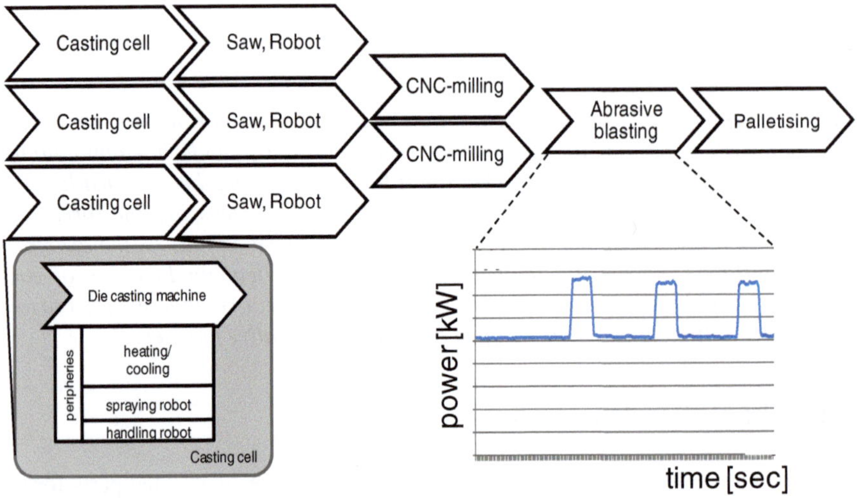

Fig. 86 Structure of considered manufacturing system

Step 4: Data Metering and Processing

In the following step relevant data was obtained for all machines. As mentioned above, it is a highly automated transfer line aiming for mass production of one specific product. Thus, all process times are well defined and can be considered as constant. Based on production respectively specifically maintenance data MTTF and MTTR were calculated which describe the failure behaviour of the machines. Finally, the electrical power demand of all consumers was measured with a portable measuring device and a sampling rate of one second. A sample result of these measurements can be seen in Figure 86. For all consumers the load profiles were analysed and consumption values were allocated to the different operation states.

Step 5: Modelling

The whole manufacturing system was modelled with the developed energy oriented manufacturing system simulation approach. Thereby, the procedures explained in Chapters 5.2 and 5.4.6 were used. As mentioned above, a casting cell contains diverse separate consumers – besides the die casting machines (DCM) these are cooling and heating units as well as spraying and handling robots. The electricity demand was measured for all of those machines. However, referring to the modelling guideline explained in Chapter 5.4.6 these consumers were integrated within one process module while their behaviour is directly connected and their integration would not limit any possibility nor lead to different results. Therewith, process time and energy consumption of the casting cell reflects the combination of these machines. Similar to this, conveyor belts that partly connect different processes automatically were not considered separately but are integrated within the related module. All other consumers were considered as separate process module whereas this most realistically reflects real circumstances and offers most flexibility for potential analysis. A screenshot of the model of the manufacturing system shows Figure 87.

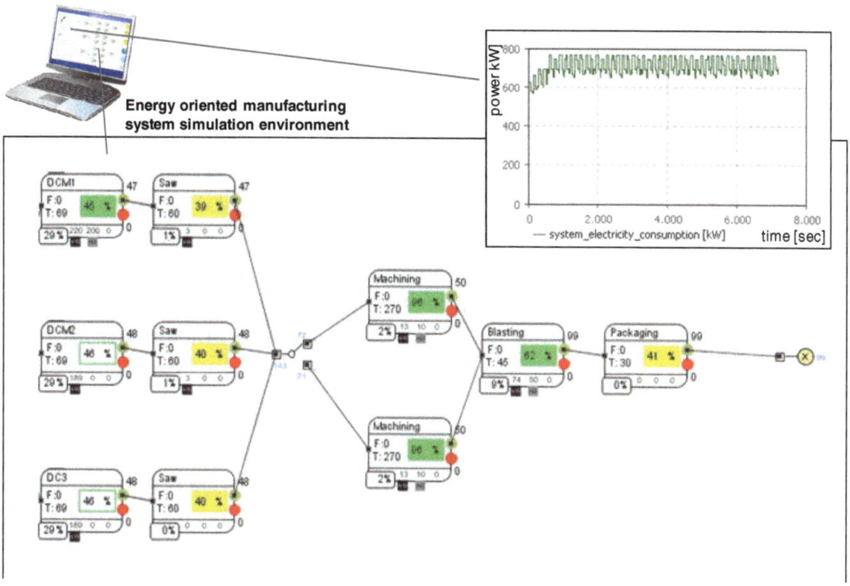

Fig. 87 Simulation model for Aluminium die casting case (results based on scenario A)

Based on the electricity measurements distinctive machine operation states could be identified that were also reflected in the model. For most machines just a distinction between "off", "idle" and "processing" is reasonable. However, the die casting machine involves a necessary spraying process after value created processing, which was considered separately.

Step 6: Validation

As mentioned before, for the considered production area no detailed energy related data was available. Therefore, validation is ensured through comparing production performance data with simulation results. For validation the maximum (nominal) production performance of all machines was collected from the company's production planning and therefore the maximum output of the system could be calculated statically. When excluding all machine failures and a quality rate of 100% in a deterministic model, similar values should be achieved through simulation ("Fixed Value Test"). For the purpose of validation one day of production was simulated and compared with the nominal output of the real system. The results are shown in Figure 88 (for reasons of confidentiality no absolute value can be given as this point).

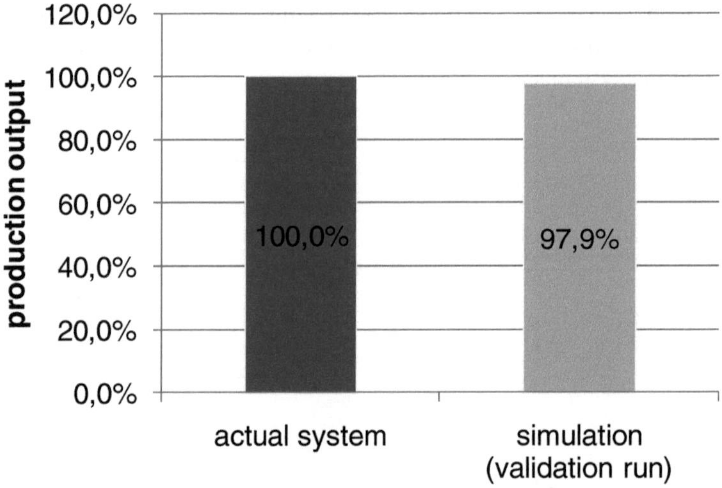

Fig. 88 Results of simulation run

The results reveal that the simulation is very capable to reflect real system behaviour. The small difference is explainable – the simulation run started with an empty system with a certain ramp-up phase until all machines were in full production. This results in less output compared to the calculated maximum production of the real system. For testing the energy consumption related behaviour a logic check based on the simulated load profile was conducted. It revealed that the simulated power demand reflected the expected behaviour and the model worked correctly. Altogether, based on these results, the simulation model of this aluminium die casting process chain can be considered as validated.

Step 7: Scenario Development / Step 8: Simulation Runs / Step 9: Evaluation

As described in the application cycle steps 7 to 9 are closely linked and part of an iterative procedure. For practical reasons these steps will be considered together in this case.

The first and obvious scenario which has to be considered is a standard scenario which depicts the real system and serves as base for later comparisons. For this purpose the values from the successful validation run were used (scenario A). Similar to the validation machine failures will be excluded at this point to avoid uncertainty and facilitate first analyses. Based on an analysis of the results improvement potentials can be identified and measures will be part of further scenarios.

The base run (see screenshot in Figure 87) reveals the high level of electrical power demand with an average value of approx. 700 kW. Due to the continuity of production (continuous flow in transfer line, no shutdown of machines when idling, no failures) demand is not strongly volatile and always stays below 800 kW. As the key figures show, main drivers of electricity consumption are the three die casting cells (which sum up for 87% of consumption), the blasting (9%) and the milling (4%) processes. From an electricity consumption perspective, the other consumers can be neglected in further considerations. The bottleneck is evidently the milling process (conducted by two machines) as the utilisation rates underline. The die casting cell has an utilisation rate of 46% which is not a result of significant idle time but rather of the necessary (but not value creating) spraying process just after each part. The blasting process is utilised with 61% which is a result of waiting for part supply from upstream processes.

The objective is to reduce energy costs (in this case directly related to electricity consumption) while still achieving comparable production output. Therefore different fields of action could be identified. Due to the nature of the system, organisational approaches are less promising since the influence of both PPC (continuous production of one product) and employees (highly automated system) is very limited. Based on the first analysis the following improvement measures will be further evaluated, which are more technical in nature and strive to decrease the level of consumption while avoiding non-value creating losses in idling times:

- The **blasting** process is not fully utilised but demands significant power even in idle mode. Therefore, a **batch production** with a lot size of 10 (instead of the current one piece flow) and an **automatic shutdown** after each batch is proposed (scenario B).
- Since the die casting machines are not the bottleneck of production it shall be analysed whether a **production with just two die casting machines** is feasible. This would dramatically decrease the power demand of the manufacturing system (scenario C).
- **Process optimisations** allow a decrease of approx. 8% of the process time **in die casting**. It shall be elaborated whether this actually has an effect from system perspective. By applying this measure the power demand of the die casting machine itself will not be reduced (scenario D).
- To reduce non-valuable idle electricity consumption, all machines with certain idle time (blasting, packaging, saw) are equipped with **systems for automated shutdown** (scenario E).
- Since it is technically possible, scenario F is a **combination of scenarios B, C and E**.

Table 22 shows the results of the simulation runs relatively to the validated base scenario. Thereby energy (electricity) consumption and production output are the target variables. However, energy efficiency is used as a key figure for evaluation. Additionally, based on 252 working days in three shift system and an assumed electricity price of 0.10 €/kWh the potential savings on electricity costs per year are depicted.

Table 22 Results of simulation runs for aluminium die casting case

Scenario	Description	Energy consumption	Production output	Energy Efficiency	yearly electricity cost savings
A	Base Scenario	100%	100%	100%	- €
B	Batch process for blasting (lot size 10) and automatic shutdown in between	92%	95%	103%	34.396,64 €
C	Just two die casting machines are used.	70%	89%	126%	127.294,12 €
D	Process time of die casting machines reduced by 8%.	100%	100%	100%	829,12 €
E	Automatic shutdown for Blasting/ Packaging/ Saw when idling.	98%	100%	103%	10.642,33 €
F	Combination B+C+E.	62%	81%	130%	162.864,83 €
*B**	*Optimised batch production for blasting.*	*93%*	*101%*	*109%*	*29.841,76 €*

Based on these results the potential impact of certain measures is proved based on the energy oriented simulation. For example, the reduction of the process time in the die casting machine has limited impact which is logical since these machines are not the bottleneck of production. Scenario E shows an improvement potential of approx. 2% regarding energy consumption while keeping the same production performance. Over the year this can total cost savings of over 10000€. As the implementation of these measures do not involve much effort in terms of time and investment (adaption of machine control or automatic shutdown system) this results in very attractive payback periods. The most interesting results are certainly those from scenarios B and C.

The batch production can reduce electricity consumption by approx. 8% and generate savings of almost 35000€ per year. As a drawback the output of production is also decreased by approx. 5% which might not be acceptable. However, the lot size of 10 for this scenario was rather randomly selected to get an impression of general potentials. With regard to the underlying potential, further analyses were conducted. Figure 89 shows the results of parameter variation simulation experiments in order to determine an optimal batch size for the blasting process (range from 1 to 20 as technically maximal capacity).

These results show that a batch size of 10 is too large as it restricts production flow. As the progression of energy consumption depicts, similar saving effects can already be generated with a smaller batch size. An optimum is found at a batch size of 4 where the production outcome is not restricted and a significant reduction of energy consumption can be realised. This is further emphasised when considering the energy efficiency value. As already shown in Table 22, applying this scenario (labelled B*) would still lead to saving of almost 30000€ per year while keeping production output constant.

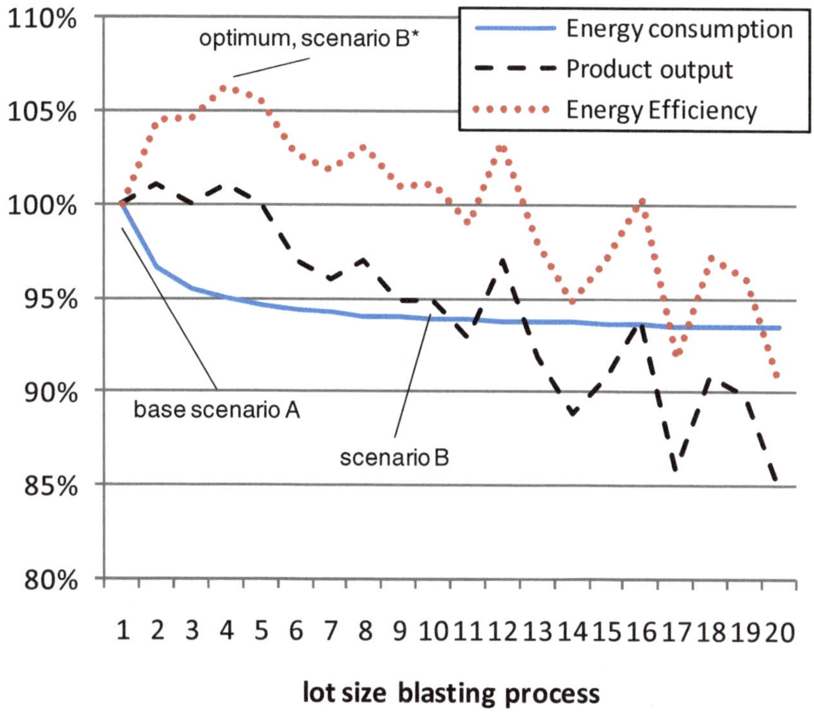

Fig. 89 Results of parameter variation experiment for batch size of blasting process

Scenario C results in a very drastic reduction of energy consumption of about 30% compared to the base run. Also the energy efficiency ratio shows very significant improvement. However, this scenario is a good example while just relying on the efficiency ratio may lead to unsatisfying results. Further criteria need to be considered and serve as necessary constrain – in this case the reduction of the production output of 11% is hardly acceptable in a mass production environment (although this decrease is relatively lower compared to energy saving potential). Nevertheless the scenario is very relevant to consider since it gives clear indications about the immense leverage of this measure, which might be an alternative if e.g. product demand is lower.

Altogether some promising means for reducing energy costs could be derived. However, all these analyses are based on a deterministic model with e.g. machine failures being disabled. While this shows the general functionality and potential there is some uncertainty regarding the outcome when assuming more realistic circumstances. Therefore extensive simulation studies were conducted which consider realistic failure behaviour of machines and shall assure applicability and potentials of the most promising scenarios B* and C. Therefore derived values for MTTR and MTTF are used and assumed as to be normally distributed. The relevance of stochastic effects transfers the former deterministic model into a probabilistic simulation model which needs a certain number of runs to be valid. There is not a single value as a result but rather a distribution of values, which has to be evaluated as a whole. Figure 90 shows the results of these simulation studies. All values are put in relation to the average outcome of scenario A. The analysis of energy consumption confirms the results of the deterministic runs and shows that significant electricity costs savings are possible with scenario B* and C. Statistical tests (e.g. t-Test) confirmed the expected effects of derived measures while differences in value distribution proved to be statistically significant (all p-values<0.05). Altogether, the probabilistic simulation runs are important to prove that measures do have the intended effect while taking into account the uncertainties in production. This can be clearly stated for scenario B* - the batch wise processing is a promising measure to reduce energy costs while keeping constant output. Compared to the existing situation, the output would be similar and an energy consumption saving potential of 8% could be confirmed. Producing with only two die casting machines (scenario C) definitely has an even higher potential from an energetic perspective, which could also be confirmed. However, probabilistic analyses even emphasise the major drawback of output reduction. This measure should only be applied if a 20-30% reduction of production output is acceptable in order to save energy costs. The range of possible values is also broader compared to other scenarios which underlines the strong dependence from failure behaviour and the strong uncertainty/risk connected with applying this measure.

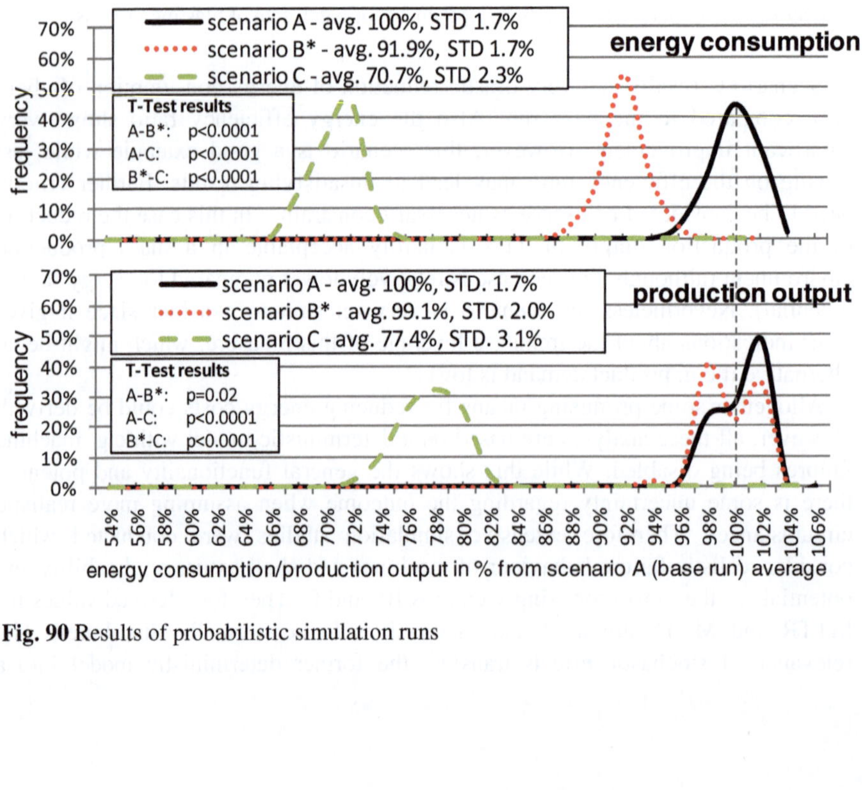

Fig. 90 Results of probabilistic simulation runs

6.2 Weaving Mill

The next case study considers a small and medium (SME) sized company running a very energy intensive and large scale weaving mill to produce technical textiles that are being used for e.g. industrial purposes (e.g. supporting material for abrasive papers, printing industry) (Herrmann et al., 2011a). The raw yarn is running through pre-treatment phases like warping and sizing before the actual weaving process is conducted on specialised weaving machines. As a last step an extensive (partly automated) inspection is conducted to ensure high quality products.

Step 1: Objective and System Definition

The company as a whole shall be considered; the objective is to minimise energy costs while still keeping up the necessary production volume.

Step 2: Total Energy Consumption and Contract Analysis

The analysis of the company's energy inputs reveals that electricity and oil are necessary to run production. As Figure 91 shows, electricity is the main energy cost driver. Due to continuous three shift production in main production areas, the electrical power demand is relatively constant on a high level (dropdowns of demand depict production pause on Sundays). Electricity costs are mainly determined by consumption related costs and standard components (fees, tax); peak surcharges play a minor role. Additionally, oil is needed for generating steam in order to run the sizing process as pre-treatment before weaving. Also warm water and space heating (of e.g. administration building) supply is ensured through oil but this is significantly less relevant compared to the demand of production.

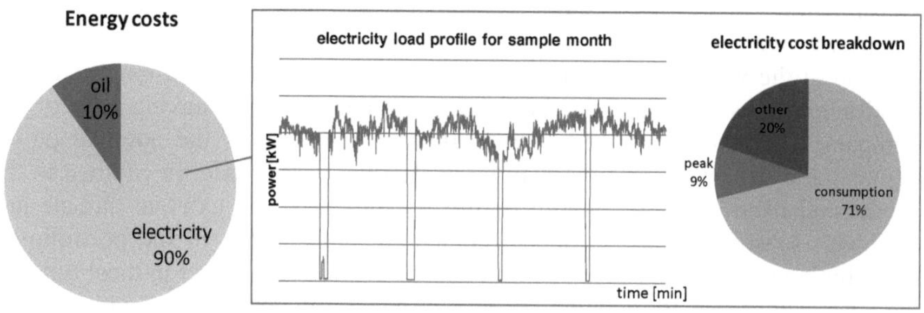

Fig. 91 Energy consumption analysis for weaving mill case

Step 3: Identification of Energy Consumers

According to the application cycle, main energy consumers shall be identified through applying the proposed energy portfolio. Since the consumers of oil are known, the portfolio was only used for the case of electricity. Therefore the nominal values for electrical power demand of different production areas and their annual operating time were derived from technical documentation and production data. Figure 92 shows the energy portfolio.

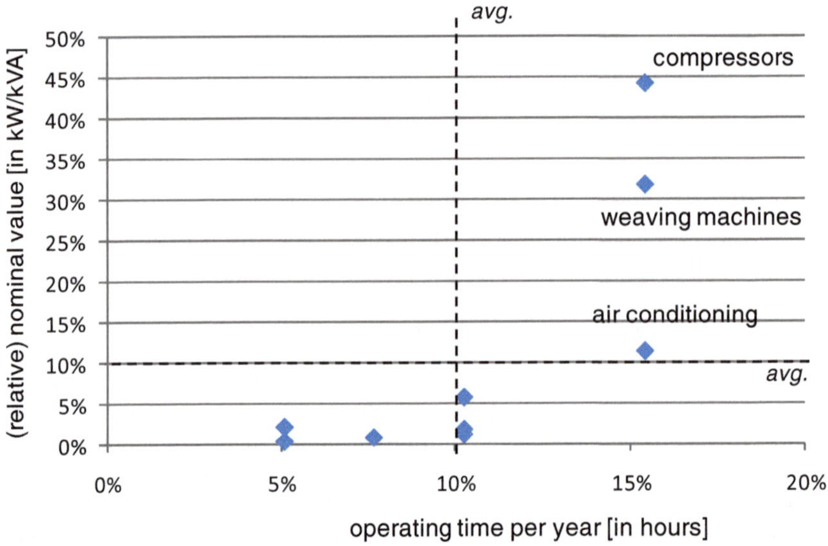

Fig. 92 Prioritisation of electricity consumers for weaving mill case

The analysis reveals that compressors and weaving machines are very likely to be by far the most relevant electricity consumers. They count up for approx. 75% of total nominal power demand and approx. 80% of estimated electricity consumption. Interestingly, both consumers are closely connected since the electricity demand of compressors is majorly induced by the compressed air demand of the weaving machines. The third largest electricity consumer is the air conditioning system which is also mainly needed for the weaving area of the company. However, on a yearly basis the consideration of the nominal power demand and the operating time may be misleading in this case. In contrast to the production itself, air conditioning is directly influenced by external climate and therewith seasonal changes. Hence, compared to the value in the portfolio its contribution to electricity consumption will be significantly lower in most times of the year. Altogether, for detailed analyses of electricity consumption, compressors and weaving machines will be considered due to their importance. All other consumers are significantly less relevant and offer minor leverage to improve energy costs.

Step 4: Data Metering and Processing

In the next step the energy consumption of the weaving machines was measured. As a result of the previous analysis it is necessary that both electricity and compressed air consumption need to be considered. The weaving mill consists of 41 machines of four different types. Consequently those four different types of machines were analysed separately with a sampling rate of one second. Additionally, the influence of different speeds was determined through varying process parameters. Sample results of those measurements are shown in Figure 93.

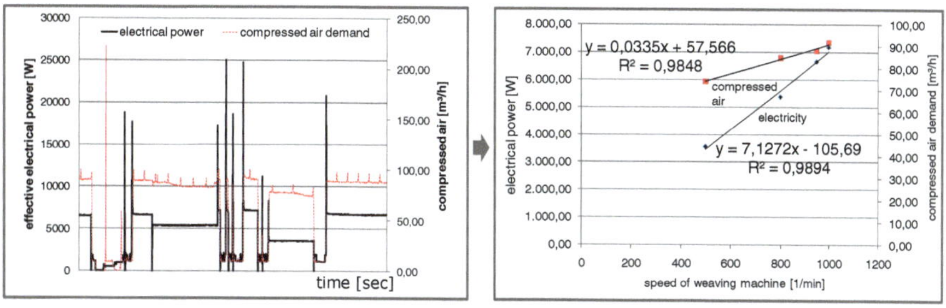

Fig. 93 Energy measurements and modelling of weaving machines

The necessary data for the compressed air system could be gathered from an online monitoring system provided by the manufacturer of the system. Based on measurements with a sampling rate of ten seconds it offers distinctive information on e.g. electrical power demand and compressed air supply as well as switching cycles of each single compressor of the system.

Step 5: Modelling

All weaving machines, the compressed air system and also the steam generation were modelled with the developed energy oriented manufacturing system simulation. The machine park of the weaving mill consists of a total of 41 weaving machines based on four different machine types operating independently (no linkage, every machine with own production program) and almost continuously in a three shift system. Based on an analysis of production data, each one of the weaving machines was modelled as one process module, with its specific characteristics on process time, failure behaviour and energy consumption. The base unit which defines “one product” needs to be kept in mind for modelling and later evaluation is one m² of textile. As shown for one example in Figure 93 for each type of weaving machine the demand on electricity and compressed air could be transferred into equations. These consumption functions reveal very good consistency with real values and allow the consideration of energy demand with respect to the speed of the machine. Additionally, both compressed air system with its seven large scale compressors and steam supply system were modelled as separate TBS modules. The necessary information regarding system structure (e.g. puffer tank size, number of compressors) and control (e.g. start-up air pressures) was derived from technical documentation and existing monitoring data.

Step 6: Validation

For validation purposes different criteria are used in order to compare results from simulation runs with real data. The validation results are shown in Figure 94 (real data=100%).

- The production output rate was derived from production data. Machine failures and setup times were excluded to ease comparability. Simulation runs with similar conditions result in very similar results with a consistency of >99%.
- Regarding electrical power demand and electricity consumption no data broken down for the considered system (weaving machines plus compressors) was available. However, based on the aforementioned estimated shares of these consumers on a company's total electricity consumption (75% of power demand, 80% of consumption) and real total values (from electricity bills) at least a rough check could be made. It turns out that simulation results fit pretty well and show the right order of magnitude.
- For the compressed air system detailed information was gathered from the online monitoring system. As base for validation, average values of selected state variables for one month of production were taken. For all values a sufficient consistency of measured and simulated results can be observed.
- To validate the calculation of the steam generation module simulated results were compared to real data regarding monthly oil demand (average value over several years). Again, a very good fit of the simulation results can be stated.

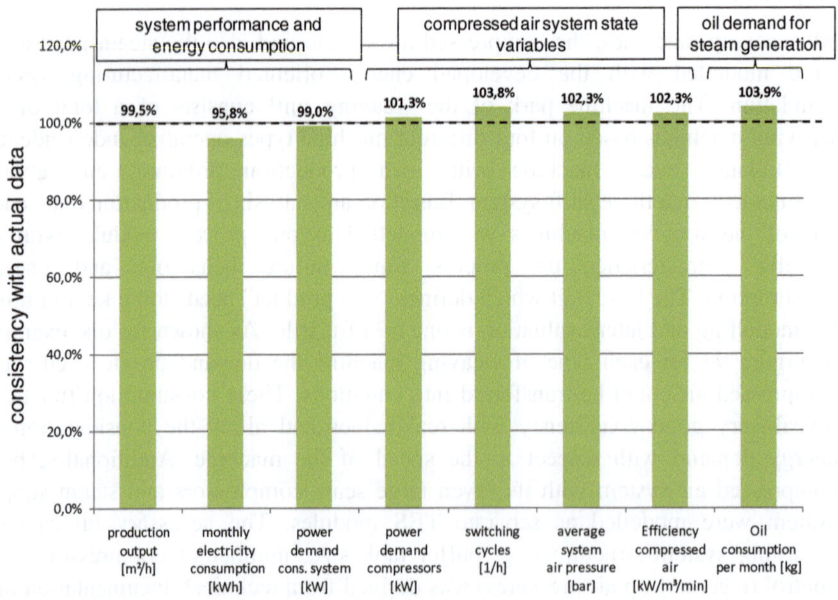

Fig. 94 Validation results for weaving mill case

Altogether the simulation achieves very satisfying results for all the considered criteria, which cover all necessary sub-systems of the company. Therewith, the simulation model can be considered as validated.

Step 7: Scenario Development / Step 8: Simulation Runs / Step 9: Evaluation

Figure 95 shows the results of the base simulation run as Sankey diagram, which are automatically generated through the E!Sankey interface. The energy flows depict the real situation in the company.

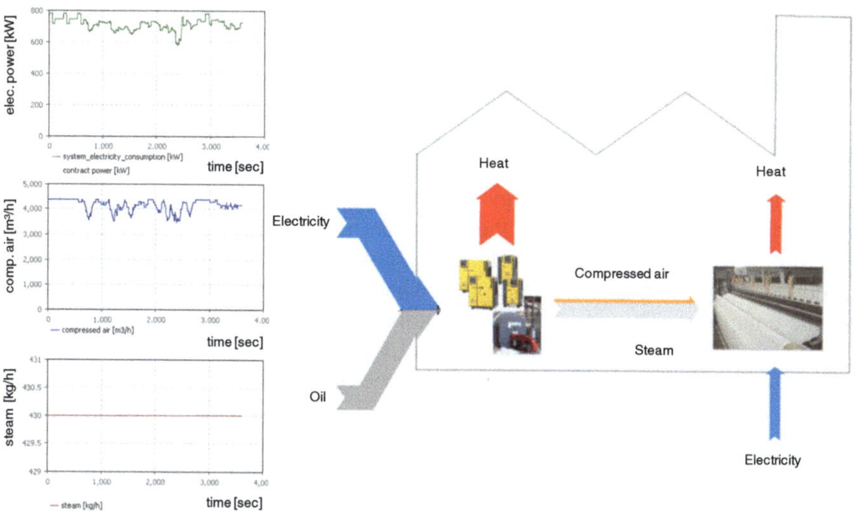

Fig. 95 Simulated load curves and automatically generated Sankey diagram of simulated energy flows (base run, in kW)

Significant amounts of electricity and oil serve as input for compressed air and steam generation. As mentioned in Chapter 2.4.2, specifically the compressors have a low efficiency, which leads to high heat losses. Only a small share of inserted electricity ends up as energy in form of compressed air. The weaving machines are the main consumer of compressed air; therewith they are directly and indirectly majorly responsible for electricity consumption. Steam is consumed in a pre-treatment process (sizing) just before weaving. However, the consumption does not depend on the behaviour of the weaving machines. Based on the results of the base simulation run (scenario A), the following fields of action shall be considered towards improvement (Table 23).

- As the measurements underline the demand on electricity and compressed air is strongly depending on the speed of the weaving machines. Therefore, the **leverage of weaving machine speed** regarding the influence on energy costs shall be analysed for the system as a whole (scenario B and C).

- A main driver of losses within compressed air systems are leakages. For the considered case the **impact of leakages** shall be determined to investigate the feasibility of actions towards leakage reduction. Pre-analysis show that a reduction by 50% could be achievable with certain effort (scenario D).
- The volume of the compressed air system is one of the determinants of system behaviour. Therefore, the impact of increasing **volume by adding puffer tanks** will be analysed (scenario E).
- Static calculations show that a stable compressed air supply of production could theoretically be achieved with **only four compressors** instead of seven. This shall be analysed through applying the simulation approach (scenarios F, G, H and I).
- As mentioned above, a significant amount of heat is occurring as a result of extensive compressed air demand. Since heat is needed to generate steam the general potential of applying a **heat recovery** system will be analysed (scenario J).

Table 23 shows the results of the simulation runs in relation to the base run. As relevant target variables of energy consumption (in kWh, electricity plus oil), the production output (in m^2) and the energy efficiency (m^2/kWh) are considered. Additionally the potential annual energy cost savings are displayed (based on production situation and market prices, which do not reflect actual price schemes of the company). The simulation runs explicitly consider failures of the machines which are - from a single machine perspective - an important influencing factor on energy demand and production output. For the base scenario A extensive analyses were conducted to evaluate the uncertainty of results against the background of stochastic occurrence of failures. Those runs reveal that the resulting values for the considered system are very robust against stochastic effects. The energy consumption has a standard deviation of only 0.5% around the average value; the production output ends up with a standard deviation of 1.3%. This can be explained through the large number of machines (minor relevance of single consumer), the continuous production situation (high share of value creating time) with the independent behaviour of the weaving machines (no linkage, so failure does not lead to failure of other machines). Thus, fluctuating failure behaviour of single machines does have limited statistical influence on key target variables from a system perspective. In further considerations, for all scenarios the same basic data set and frame conditions compared to the base run are assumed (e.g. failures will occur in a similar manner). Hence, comparisons are possible and also statistically legitimate due to a high level of statistical certainty.

Scenarios B and C assume a general (all weaving machines) reduction respectively increase of machine speed by 10%. Not surprisingly energy consumption and production output are directly influenced. However, a 10% change of speed led to just 2% change of energy consumption (direct and compressed air induced) whereas the effect on production output is 9-11%. Motivated by the outcomes of these scenarios more detailed runs were conducted. The results are shown in Figure 96 and consolidate the results. The leverage on energy consumption is relatively low while production output is heavily influenced. Thus, from an energy

Table 23 Results of simulation runs for weaving mill case

Scenario	Description	Energy consumption	Production output	Energy Efficiency	yearly energy cost savings
A	Base Scenario	100%	100,0%	100%	- €
B	Reducing production speed by 10%.	98%	91,0%	93%	15.247,87 €
C	Increasing production speed by 10%.	102%	111,3%	109%	-18.081,79 €
D	Leakage is reduced by 50%	98%	99,5%	102%	18.724,61 €
E	Increasing puffer tank size. (doubled)	100%	99,6%	100%	2.965,25 €
F	Just compressor 1-4 used.	86%	52,0%	60%	112.444,42 €
G	Just compressor 1-4 used+doubled puffer	92%	74,7%	81%	64.903,68 €
H	Just compressor 1-4 used+reduced leakage	96%	99,7%	104%	32.908,03 €
I	Just compressor 1-5 used	100%	99,4%	100%	2.592,00 €
J	Heat recovery	76%	100,0%	132%	73.345,54 €
K	Combination of scenarios H and J	74%	99,7%	135%	99.479,81 €

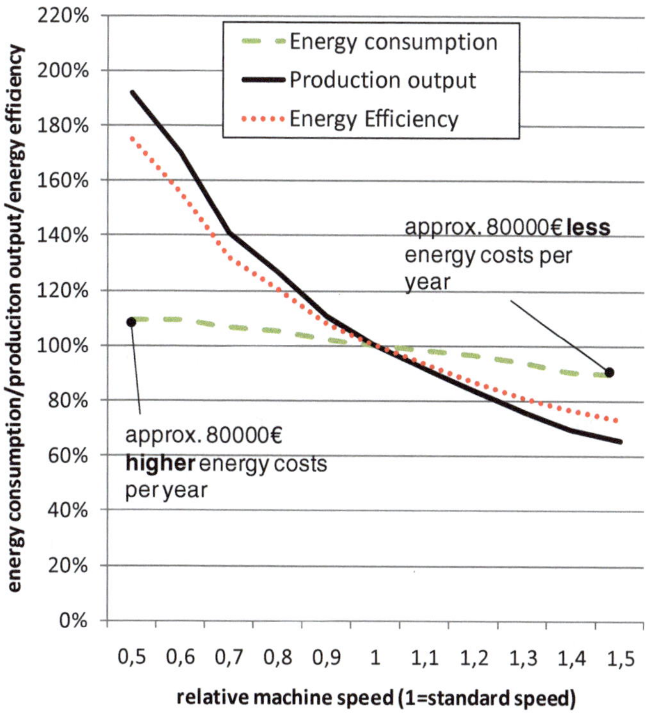

Fig. 96 Impact of changing speed of weaving machines

efficiency perspective operating as fast as possible is the most promising strategy. However, it has to be mentioned that coherences of failures probabilities and machine speed are not considered here due to lacking data. Significantly faster (or also slower) production is likely to result in more failures. Still the impact on energy costs per year is very considerable with an average of approx. 16000€ per 10% speed change. While certainly being no strategy in times of economic upturn (due to less output), reducing machine speed could be a useful strategy for low utilisation phases and is a way to bring down operating costs.

With the next scenarios a closer look on the highly relevant compressed air system is taken. Scenario D reveals that reducing leaks bears a significant cost reduction potential, which sets the limit for the necessary efforts in terms of time and costs/investment to achieve those reductions. Increasing the puffer tank size has limited effect on energy consumption and production output (scenario E). However, analysing further technical variables shows that it reduces the number of cycling switches of compressors by 50% which may extend their lifetime. Based on those results further investment calculations regarding the feasibility of integrating additional tanks are conducted.

Having in mind the measured and simulated compressed air demand of weaving machines it seems possible to significantly reduce the necessary number of compressors. This would bear an enormous saving potential in terms of energy and other operating costs and investment. While this would theoretically cover weaving machines' demand, in the first step, the operation of only the largest four of the existing compressors was assumed (scenario F). This option turns out to be not feasible for practical application with the given circumstances. The simulation reveals that the necessary air pressure of the system would drop down too much, causing failures of the weaving machines and enormous output losses. With the simulated time depending consumption behaviour and assumed leaks in the system, four compressors are not sufficient for a stable production. To ease the problem of dynamically fluctuating demand with some peaks, the puffer was doubled in the next step (scenario G). This can in fact improve the situation – however, the result in terms of production output is still not acceptable. While leakage of the system is evidently a significant factor, similar to scenario D a reduction of 50% was assumed in scenario H. In fact, this reduction in constant compressed air demand enables a stable production with just four compressors. This gives further incentives to focus on reducing leaks in the compressed air system. Without reducing leaks or increasing puffer size a stable production is possible in any case with five compressors. However, from an energy saving perspective, just minor effects can be expected, which underlines that the additional two compressors of the existing system (and base run) are rarely operating anyway. Still this information is very important for further investment decisions. Finally, the base run reveals the immense heat potential, which is available due to compressor operation while the neighbouring boiler needs heat to generate steam. Therefore, the potential of applying heat recovery was analysed (scenario J). With a given efficiency (oriented at state of the art heat exchangers) the exhaust heat of the compressors can serve as energy input for steam generation. As the results show a significant reduction of additional energy demand can be achieved. In contrast to the other scenarios the energy cost reduction is a result of less oil needed. While this is just a rough estimation and detailed technical analyses are necessary, cost savings of >70000€ are possible, which justifies the investment in necessary heat recovery technology. Finally, scenario K combines the promising scenarios H (only four compressors used with 50% reduction of leakage losses) and J (heat recovery) to evaluate the overall impact. This is a very good example that underlines the necessity of a holistic and simulation based evaluation of measures to be able to analyse the impact of their interactions. In this case both measures counteract to some extent. Still the saving potential is very promising even though it is not the sum of the possible savings of the isolated consideration of those scenarios: the reduction of compressor power demand directly leads to significant energy savings. However, also less waste heat as source for heat recovery is available. Thus, less oil can be saved compared to the original heat recovery scenario J.

6.3 PCB Assembly

The third case study considers a SME company which assembles printed circuit boards in a very flexible production environment (no coupling of machines, free flow of orders). Main processes are the mounting of parts on PCBs with SMD (surface-mounted device) or THT (through-hole technology) techniques, which involve reflow soldering machines/ovens (SMD) and solder waves (THT).

Step 1: Objective and System Definition

The company as a whole shall be considered; the objective is to minimise energy costs while still keeping up the necessary production volume.

Step 2: Total Energy Consumption and Contract Analysis

In this case the only externally acquired form of energy is electricity. Figure 97 shows an analysis of the daily electrical power demand as measured by the electricity supplier with the usual sampling rate of 15 min.

It is a statistical analysis from 60 working days meaning that the value of each depicted 15min period is a result of 60 single values (e.g. average power demand at 10 a.m. is the average of 60 different days at this point of time). It is important to note in this context that this analysis can just provide general tendencies regarding consumption behaviour – the company consists of a highly flexible production environment so basically each day can look quite unique. The consumption pattern of a typical working day is not as even as the average curve. Also minimum and maximum values are not representative for a specific day but provide a corridor to describe the most likely power demand pattern. This corridor can be narrowed down, by considering the average plus respectively, minus standard deviation – per definition this range contains 68.2% of all values. In general it can be stated that coming from an average base load of around 4 kW, production typically starts at around 6 a.m. with switching on diverse machines,

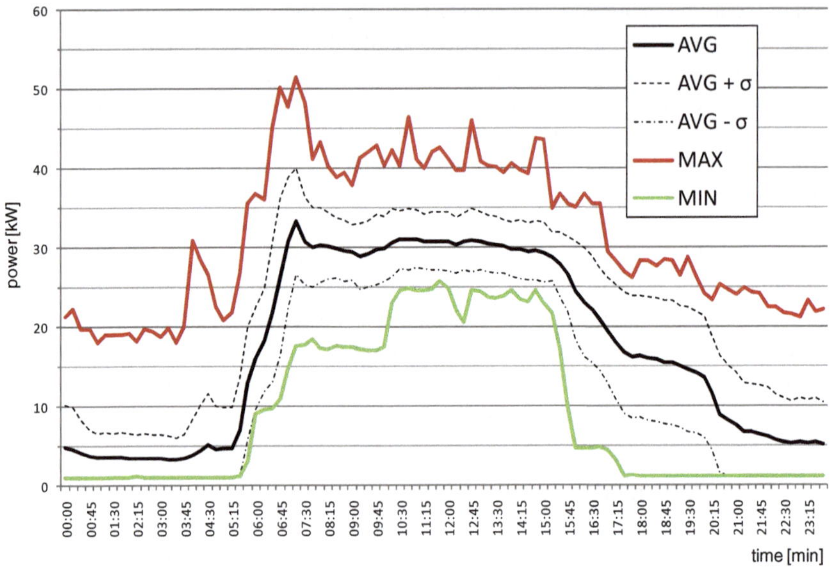

Fig. 97 Electrical power demand of PCB assembling company

which leads to a peak demand on electrical power. After 2-3 hours, consumption is relatively constant on a certain level. After 4 p.m. a significant dropdown of power demand can be observed which is related to the typical daily shift ending. Sometimes selected production machines are operating longer to cover necessary production volumes.

In the specific case of this company electricity, costs are directly related to consumption with one constant price rate for the whole day. There are no peak surcharges.

Step 3: Identification of Energy Consumers

Figure 98 shows the energy portfolio of the considered company with respect to nominal electrical load and estimated operating time per year. It can be observed that electricity consumption is heavily determined by four main consumers which sum up to over 87% of the estimated yearly consumption: two solder waves, one reflow oven and the compressor for compressed air generation. From an energy point of view all other consumers are of minor relevance.

Step 4: Data Metering and Processing

Based on the outcomes of the previous step, measurements of electricity consumption were conducted for diverse consumers. Figure 99 shows an example measurement for the reflow oven (sampling rate of one second). The main consumers involve extensive time and related electricity consumption for heating up, which is necessary prior to any value creating activity at each working day. Based on measured data, a comprehensive check was conducted in order to confirm results of the aforementioned portfolio and ensure that all main consumers were actually considered. The results of this check are also displayed in Figure 99. To ensure the comparability the maximum load within a 15 minute interval – as base for electricity consumption measurements done by the supplier - was calculated for each consumer (based on measurements with sampling rate of just one second). The comparison of these values with the power demand of the company as a whole (Figure 97) confirms the previous considerations. The typical demand (average and standard deviation range) can be well explained through these consumers; the higher absolute maximum is a one-time value and is most likely the result of incidentally switching on the second reflow oven.

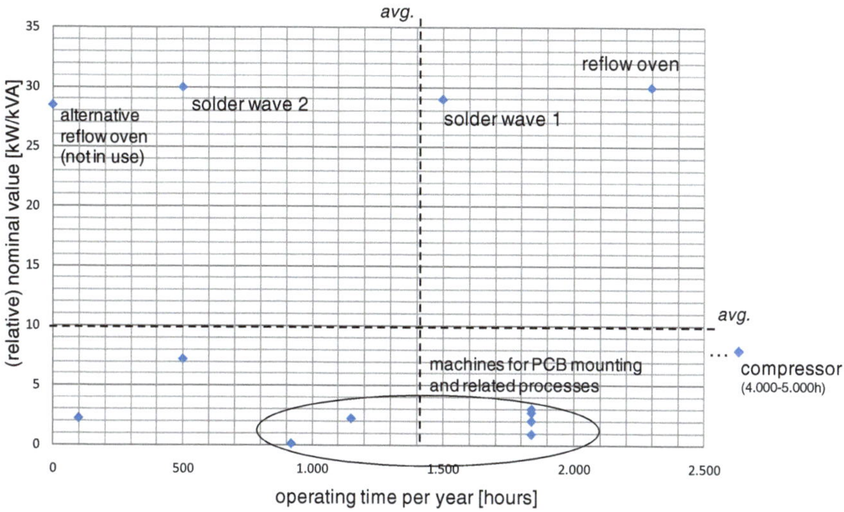

Fig. 98 Energy portfolio of PCB assembling company

Step 5: Modelling

In contrast to the other case studies, the generic modelling approach was used for this case. As presented in Chapter 5.3.1 and even more specifically in Chapter 5.3.5 it is possible to setup the whole model via parameterisation of generic process (and also TBS) modules "outside" the actual simulation environment in a standard MS Excel™ datasheet (PPC module). This addresses entry obstacles and eases utilisation of simulation capabilities without having the necessary detailed expertise. The considered company is a typical SME which does not have the necessary experts and financial possibilities to run complex simulation tools (whereas commercial energy oriented simulation approaches are not available anyway as shown above).

All necessary data was keyed in the PPC module. Thereby, all relevant consumers were modelled according to the proposed procedure depicted in Figure 83. Thereby it is necessary to mention that - in accordance to the proposed procedure - certain machines were modelled as separate process module although they are per se not relevant from energy perspective. However, these machines are crucial to consider as they determine the material flow within the company which again influences the operation of main energy consumers. In total, seven different production machines were modelled (reflow oven, two solder waves, two devices for PCB mounting, automated optical quality inspection, machine for automatic application of soldering paste). Based on the data analysis an electricity base load was modelled subsuming minor consumers, which do not need to be considered separately (according to Figure 83). As mentioned before production machines are not technically linked with each other and operate independently. The material flow is determined by production planning of orders that run through the company in flexible process chains utilising the given production equipment e.g. in diverse sequences and lot sizes. Based on realistic circumstances, certain general process chains were modelled via the PPC module for three different types of orders (products) and a typical production program for a working day (involved machines and sequence for each order, lot sizes, starting time) was developed. Supplementing the production equipment and management, the compressor was modelled as TBS module based on its technical specifications. Compressed air is mainly used by PCB mounting devices. Their compressed air demand was estimated based on nominal values.

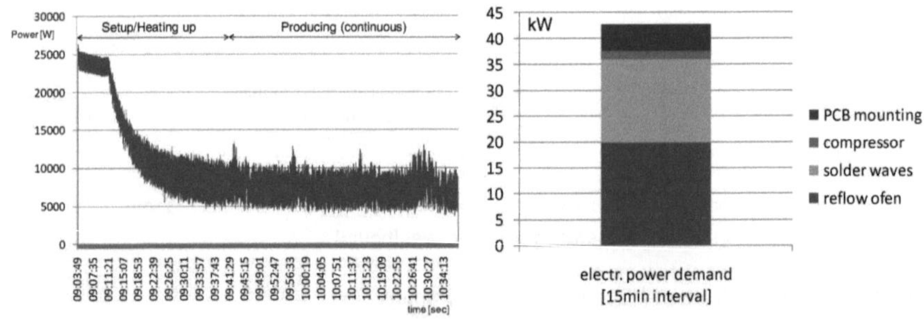

Fig. 99 Example measurement result of reflow oven and cumulated maximum power demand in 15 minute interval for main consumers

Step 6: Validation

Due to the aforementioned flexible production environment in combination with lacking production data the validation is conducted through assessing the energy related behaviour of the system. If this turns out to be realistic the production performance as a whole can also be considered as sufficiently realistic since product individual setup and process times were actually measured. Figure 100 shows the simulated electrical load profile (calculated with 15 minute interval) for the validation run, which considers a typical working day with production start at 6a.m. and consequent shift end at 4p.m. A comparison with the given real electrical load profile intuitively shows the very similar consumption behaviour over the day. This can also be expressed quantitatively: with the given production scenario the validation run results in an energy consumption of 363 kWh over the considered period of time. For the same period the analysis of real data reveals an average consumption of 312 kWh with a standard deviation of 51 kWh (261kW-363kWh). Therewith the simulated consumption reflects the real consumption behaviour while even being in the range of the most likely values. An alternative scenario (B, later explained in detail) with a slightly modified machine start in the morning results in a less distinctive power peak while leading to a consumption of 357 kWh. Altogether, it could be proved that the simulation model is capable to describe the consumption pattern of the PCB company, with sufficient accuracy and can be considered as validated.

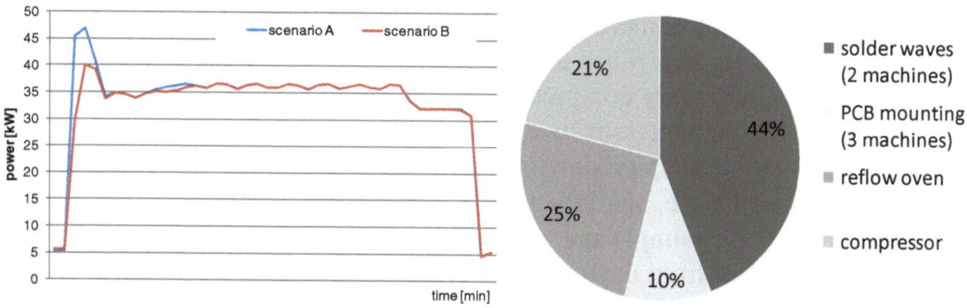

Fig. 100 Simulated electrical load profile for PCB case (second based values converted to 15min interval) and consumption composition for scenario A (base scenario)

Step 7: Scenario Development / Step 8: Simulation Runs / Step 9: Evaluation

The simulation run for validation purposes helps to understand the energy consumption dynamics of the company. As a result of this base run, Figure 100 shows the composition of energy consumers. The distinctive power peak in the morning is a result of turning on and heating up soldering waves and the reflow oven. Similar to the real load profile there is also a dropdown observable after the reflow oven reaches the necessary temperature for operation. There is again a slight increase when the soldering waves start operation. In the base scenario all machines are on for the whole shift and are turned off at shift end (4p.m.), which seems to be a realistic assumption when comparing to real values. However, they are not necessarily completely utilised for value creation due to involved heating periods and waiting for upstream supply of parts (utilisation rate of 10-20%).

Against the background of these observations, different scenarios are derived in order to improve the energy cost situation of the company. Besides technical measures organisational aspects are of specific interest here due to the flexible production environment and possible influence of machine operators. Additionally, a closer look shall be taken on the energy contract situation of the company. An overview of the considered scenarios and respective results is given in Table 24. All scenarios were simulated with a deterministic model while uncertainty is a minor issue in this case (very high availability of machines, clearly determinable process times).

- Scenarios B-E focuses on alternative strategies for production planning and control (PPC) the activities of the manufacturing system. As described above within the base run, production machines are turned on and off at shift start and end independently from their utilisation. Scenarios B-E aim at a more efficient operation of machines with reducing waste energy in idling times through purpose oriented machine start and e.g. more batched production. In scenario B each machine is firstly turned on (including heating) when the first part from the upstream process is arriving. As further extension scenarios C-E introduce an energy oriented production program (capacity) planning with distinctive starting times for the critical machines while aiming at still providing the necessary output.
- The generation of compressed air also has a relevance regarding the energy costs of the company. Since the existing compressor is dimensioned relatively large for the necessary compressed air, in scenario F it was analysed whether an alternative compressor with lower air supply, but also lower power demand could keep up production and save energy costs.
- As described before, the structure of the current electricity contract of the company is rather simple and only includes a single consumption related price rate independent from daytime. In scenario G it shall be investigated whether an alternative contract setup could lead to energy cost savings. Therefore energy costs of the base scenario were calculated with the contract from a comparable SME company. This contract consists of e.g. peak power surcharges, which allow a lower price rate for consumption.

Table 24 Simulation results overview for PCB company case

Scenario	Description	Energy consumption	Energy costs	Production output	Efficiency
A	Base Scenario	100%	100%	100%	100%
B	Machines are turned on if first upstream part available (PPC1).	98%	98%	100%	102%
C	Energy oriented production planning (PPC2)	60%	60%	77%	128%
D	Energy oriented production planning (PPC3)	68%	68%	100%	148%
E	Energy oriented production planning (PPC4)	66%	66%	100%	151%
F	alternative compressor	98%	98%	100%	102%
G	alternative electricity supply contract	100%	119%	100%	100%

The simulation results reveal the significant potential of energy sensitive production planning and control. For the considered case some PPC strategies offer saving potentials of 30-40% compared to the base scenario. As an example, Figure 101 shows the load profile of the base scenario A and scenario E. However, there is evidently also the possibility of negative impact on production output (scenario C), which is usually not an option in industrial practice. Additionally, these measures require the necessary flexibility with a certain degree of freedom in daily operation. This is not given in high volume production with rigid linkage like e.g. in case study one (die casting). However, in the considered case there is this degree of freedom. Production orders do not require the full daily capacity of available machines and operators are flexible to be appointed (can be involved in other activities within the company, flexible time contracts). Against this background scenario E is a promising way to gain energy efficiency. Instead of continuous running but often idling energy intensive production machines over the whole day, certain production volumes are batched and processed in the afternoon. The simulation confirms that the same production output can be achieved. Operators are free to be appointed to other processes in the morning or can start their shift later.

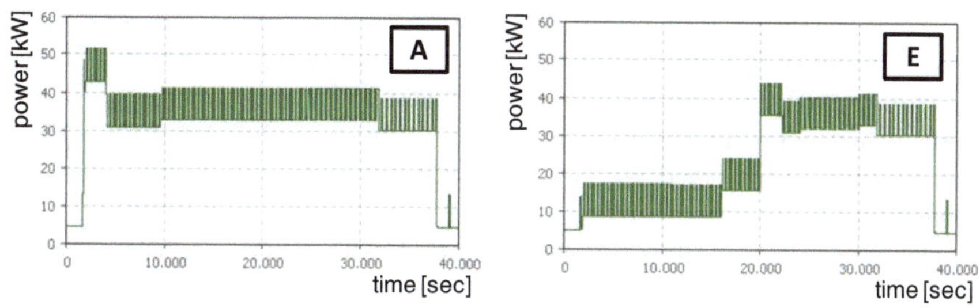

Fig. 101 Selected simulated electrical load profiles

Certainly, the scenarios only cover a sample day with selected production orders and involved process chains; the effect of those measures needs to be verified for differing production scenarios and in industrial practice. However, the general potential is promising and the achievable order of magnitude encourages the application of energy sensitive planning and control in the company. Since just selected scenarios were considered, through a more systematic design of measures even more saving potential seems to be possible for certain cases. The ultimate goal should be to include those considerations in daily planning procedures, e.g. within ERP systems.

As a more technical measure, an alternative compressor was used for compressed air generation (scenario F). The simulation proved that the smaller sized compressor is still able to sufficiently supply the manufacturing system while leading to some savings in energy consumption.

Finally, the alternative contract model is not appropriate when being applied on base scenario A. While reducing the consumption costs, the influence of the power induced cost component is highly significant and weighs more. Still it could be proven that alternative contract models can be included in the simulation approach and may have significant impact on occurring energy costs.

6.4 Application in Education of Production Engineers

Besides the application in industry towards an improvement of energy efficiency, the proposed simulation approach was already used in the education of engineers. Since autumn 2009 the Product- and Life-Cycle-Management Research Group within the Institute of Machine Tools and Production Technology (IWF) of TU Braunschweig offers an innovative lecture named "Sustainability in Production Engineering" (Herrmann et al., 2010b). Therewith students of mechanical and industrial engineering shall be sensitised and enabled to handle issues like energy efficiency in manufacturing. From a longer term perspective this is a very promising approach since these students will be the relevant experts to plan, evaluate and decide on those issues in companies later on. Thus, there is a strong leverage by motivating them and providing necessary methods and tools. The lecture pursues a practical approach: besides the theoretical background provided in the lecture itself, laboratory classes and further tutorials are offered. In groups of approx. five people, the students independently solve a comprehensive case study, which aims at increasing the energy efficiency of a defined machining process chain. They practically need to apply knowledge and experience related challenges.

The task of the case study is to produce a shaft with defined geometry as energy efficient as possible. In the first step the students take a closer look at necessary processes. For achieving the necessary final diameter and surface quality (fine machining) as the main production task, with turning or grinding two process alternatives are available. Within the laboratory environment of the institute, the students conduct themselves measurements on machines to assess process times and electrical power demand. With these results process recommendations for the most energy efficient way of manufacturing can be given. In the next step, the process chain as a whole is considered which includes several working steps – rough turning, hardening, fine machining and cleaning. The task is to produce 100 shafts as energy efficient as possible. For assessing the technical performance and energy consumption of the process chain, the proposed energy oriented simulation approach is used. For facilitating access in terms of training time and necessary software installations, the possibility of AnyLogic™ of generating an executable Java applet was used. Therewith it is possible to provide a generic model of the process chain which can be run in any web browser. The model with its graphical user interface is depicted in Figure 102.

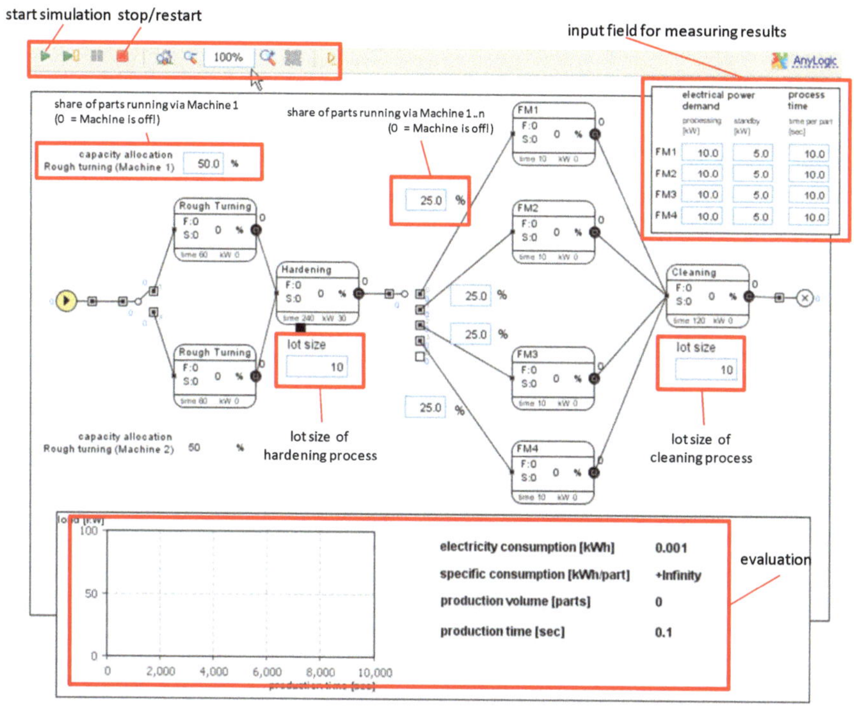

Fig. 102 Screenshot of Java-applet for energy oriented manufacturing system simulation for educational purposes

It provides the general framework of the considered process chain while leaving sufficient freedom to test alternative means to structure and control the system.

- Number and type (turning, grinding) of processes and the allocation of production volume shares to single machines can be changed.
- Measured data for processes can be integrated; therewith the effect of process alternatives can be analysed from process chain perspective.
- Lot sizes for hardening and cleaning can be changed.

As formerly mentioned the objective is to produce 100 shafts as energy efficient as possible (in accordance with the measurements, only electricity is considered here). The presented influencing options lead to significantly different results in terms of energy consumption and production output, which need to be balanced. Therewith the simulation model can strongly support the comprehension of interactions in the process chain and possible conflicts of goals. It was successfully applied in the education of engineering students and the feedback was very positive. In future the simulation based case study will be developed further and continuously used for teaching.

Chapter 7
Summary and Outlook

In the final chapter, the contents in general and the research outcomes in particular will be briefly summarised. Additionally, an evaluation of the developed concept will be conducted and an outlook on further possible work will be given.

7.1 Summary

This book aims at contributing towards improving energy efficiency in manufacturing through developing a generic energy flow oriented manufacturing simulation environment. The environmental and economic motivation is outlined in **Chapter 1**, which also stresses the necessary consideration of all dimensions of sustainability as new paradigm in manufacturing.

In **Chapter 2** the relevant **theoretical background** based on the known state of the art is described. First of all, necessary definitions and basics in the field of production and energy (supply) are given. Afterwards both aspects are brought together when describing the background of energy consumption and energy efficiency in manufacturing. It is pointed out that the topic in general is certainly not new and good basic knowledge exists in literature. However, the reasoning also underlines that implementation in industry is still unsatisfying and worthwhile potentials can still be tapped. The description of the theoretical background also stresses the diversity of questions, challenges and disciplines which are connected with the topic of energy efficiency in manufacturing.

Taking this into account, **Chapter 3** considers the **derivation of general requirements** from both scientific/technical as well as industrial/business perspective, which should be addressed to successfully foster energy efficiency in manufacturing. Thereby the discussion reveals that considering the interacting dynamic consumption profiles of all relevant energy flows for the factory system as a whole is crucial for the evaluation of improvement measures on both single machine as well as factory level. To assess those consumption profiles with static calculations, fuzzy logic, artificial neural networks and manufacturing system simulation, different methodological approaches are available. The comparison with the requirements clearly shows that manufacturing system simulation is the most appropriate methodological approach for the considered research task.

S. Thiede: Energy Efficiency in Manufacturing Systems, SPLCEM, pp. 171–177.
springerlink.com

Whereas not available in industrial practice, in research, first approaches towards energy oriented manufacturing system simulation were developed. The requirements were further concretised into specific criteria, which serve as a basis for the evaluation of the **state of research** in **Chapter 4**. It turns out that there is significant room for improvement in all areas and there is no approach with a balanced and high fulfilment of all criteria.

This evaluation shows the necessary research demand, which sets the frame conditions and necessary specifications of the **concept development** being conducted in **Chapter 5**. The core element is an innovative energy oriented manufacturing simulation approach. It is a one-stop solution with modular and flexible structure that aims at facilitating access to simulation capabilities for broad fields of application and provides decision support towards a systematic improvement of energy efficiency. Thereby it is explicitly able to consider all relevant forms of energy needed by production equipment or technical building services as well as their interactions. The developed simulation environment is embedded within an application cycle, which supports the goal-oriented implementation for the derivation and evaluation of improvement measures. Even without usage of the simulation it is a helpful methodology to identify and prioritise main energy consumers in manufacturing companies.

Finally, in **Chapter 6 (concept application)** the developed simulation based concept as a whole is applied to a couple of very different case studies in order to demonstrate the broad applicability and potentials. The case studies show that the concept is working for diverse practical applications (large and SME companies of different branches, engineering education) and underlines the diversity of energy efficiency measures which can be evaluated on a realistic base.

7.2 Concept Evaluation

Finally, the developed simulation based concept for improving the energy efficiency in manufacturing shall be evaluated. To support a structured and objective assessment the same criteria as for evaluating the state of research are used (Chapter 4.1). The results are shown in Table 25.

The evaluation shows that all criteria could be addressed and improvement towards the state of research could be made in all categories. The average fulfilment of criteria is 0.88 (88%) which is by far the highest value for all comparable approaches. This is certainly not too surprising since these criteria served as requirements for the development of the proposed energy oriented simulation approach. However, it shows that these identified critical aspects could be successfully implemented. Figure 103 shows a more detailed comparison with the state of research and underlines that significant advances are made in all areas.

Table 25 Evaluation of proposed simulation approach

Main Author(s)	Sebastian Thiede	
Institutions	TU Braunschweig, Institute of Machine Tools and Production Technology, Product- and Life-Cycle-Management Research Group, Braunschweig, Germany.	
References	(Herrmann and Thiede, 2009a, Thiede and Herrmann, 2010, Herrmann and Thiede, 2009b, Herrmann and Thiede, 2008; Herrmann et al., 2011a)	
Software tools	AnyLogic, MS Excel (parameter input, extended evaluation), E!Sankey (Energy flow visualisation)	
Energy and Resource Flows		
Completeness	●	all external energy sources and internal flows can be considered
Dynamics	●	state based consumption of all energy flows, cumulative profile available
TBS	●	interaction with two TBS systems already implemented, extension possible
Fields of action		
Technological	●	basically all technical fields of action can be addressed, also from integrated perspective
Organisational	●	diverse PPS and employee behaviour related aspects can be addressed
Optimisation	◕	Optimisation studies are possible through OptQuest™-solver; however, not main focus and just exemplarily conducted
Evaluation		
Economic	◕	costs of several energy carriers with realistic contract based cost model for electricity, no extended cost model with other cost portions
Ecological	◑	conversion of different energy flows to GHG emissions
Technical	◕	output/production rate and further key figures (e.g. utilisation) available

Table 25 (*continued*)

Decision Support	◕	diverse key figures and methods for integrated evaluation are introduced, from scientific perspective extension towards MCDM possible
Uncertainty	●	is addressed and appropriate methods for coping with uncertainty are presented
Implementation		
Transferability	●	highly transferable to diverse production cases, eased access also for non-expert
Effort	◕	little expertise necessary, fast modelling possible, possible export function as java applet, simulation time depending on actual computer - complex models might involve significant computing time
Visualisation	●	material flow/machine states and all results continuously displayed during runtime
Application Cycle	●	comprehensive application cycle provide which guides user through usage towards systematic improvement

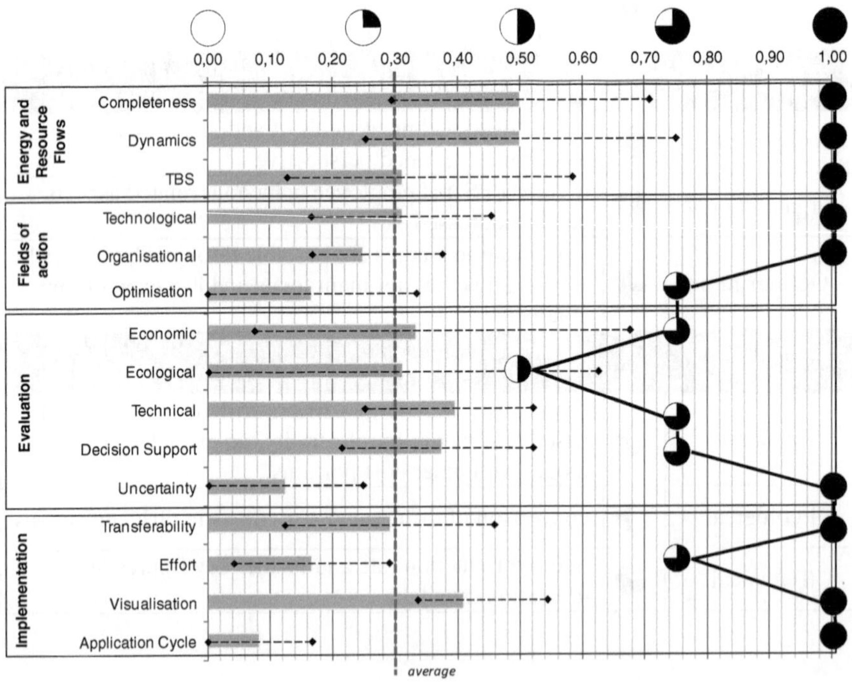

Fig. 103 Comparison of proposed simulation based concept with state of research

The comparison reveals the success factors of the proposed simulation approach compared to the state of research which are:

- The consideration of the manufacturing system as a whole with all relevant energy flows and their time dependent dynamics.
- The applicability for a broad range of production cases (including flexible manufacturing systems) of different branches with relatively low effort and necessary expertise.
- The integrative derivation and evaluation of diverse improvement measures from different fields of action.
- The provision of clear decision support through meaningful key figures and diagrams while also addressing the issue of uncertainty.
- The integration within a comprehensive application cycle, which guides also non-experienced users through a systematic improvement process while using the simulation approach.

The simulation based concept was developed and practically implemented; applicability and potentials were successfully proved through application to several very different case studies.

7.3 Outlook

Although significant advances in comparison to the state of research could be made there are diverse opportunities for future research, which would extend functionalities of the developed approach and support embedment in industrial business processes.

- **Extension and addition of TBS modules:** Within this book the development of compressed air and steam generation as TBS modules is presented, because they are not only of main relevance in general but also particularly for the case studies. These modules were verified and can be easily parameterised in order to depict any supply systems in manufacturing companies. However, further generic TBS modules are worth to be developed. As one example *air conditioning systems* are an interesting field of action while being relevant for many companies and often a major consumer of energy. Also the integration of electricity supply systems is a promising approach. Being neglected so far, the *local generation of electricity* certainly bears potential to foster sustainability in manufacturing specifically when it comes to renewable energy sources (e.g. solar or wind power). Through simulation of electricity demand of production the developed approach can strongly support dimensioning and control of those systems. Again, this is specifically of interest when different forms of energy act together, e.g. in combined heat and power generation (CHP). Additionally, further TBS modules are certainly worth to be developed for *decentralized cooling respectively heating devices* for production machines (e.g. cooling for laser cells, process heat for casting). These devices are widely distributed in manufacturing companies and so the provision of generic TBS module would be very helpful.

- **Consideration of further resource flows:** This book focuses on the development of an energy flow oriented simulation of manufacturing systems. The *consumption and emission of other resources (e.g. auxiliary materials, waste streams)* is only partly considered (as long it is relevant for energy related considerations). Technically an implementation is not too difficult due to the flexible structure of the provided simulation environment. Some of those variables are already integrated in e.g. process modules but – for reasons of comprehensibility and the actual definition of the designated research task – were not put in the main focus of the considerations here. However, in further research, consumption and emission of additional resources should be integrated while this provides a further perspective on sustainability in manufacturing.
- **Extended cost models:** The embedded cost evaluation concentrates on the realistic evaluation of energy costs. Further cost portions are not explicitly in the simulation itself and need to be calculated separately. In fact it would make sense to integrate those issues in future. Through the extension with other resource flows (see above) and worker operations, a *comprehensive calculation of total operating costs* is possible. Even more, through integrating more basic data (on e.g. necessary investment, financial conditions or maintenance strategies) even a calculation of *life cycle costs* is basically thinkable. Altogether extended cost models would support the decisions towards energy efficiency measures but also broaden the perspective to evaluate any kind of changes in the manufacturing system.
- **Extension of environmental evaluation towards Life Cycle Assessment (LCA):** While up to now an energy oriented conversion into GHG emissions is provided for environmental evaluation, an extension with other relevant resource consumption and emission patterns would enable a more comprehensive assessment of environmental impact. Through interfaces with life cycle inventory databases (e.g. EcoInvent™) even an *automatic life cycle assessment (LCA)* as ultimate goal is possible.
- **Integration of complex operations research (OR) and statistical methods:** Methods for optimisation, multi-criteria decision making and evaluation under uncertainty are an integral part of operations research as business related research discipline. Very complex methods are available in this context and can be (manually) applied in context of the data which is generated by the developed simulation environment. However, in this book mostly simplified approaches are used in order to facilitate practical application. To improve the quality of decisions, the integration of proper methods (e.g. optimisation algorithms for optimal production capacity or lot size planning, guided PROMETHEE evaluation, automated statistical treatment of stochastic simulation data) as an inherent part into the simulation is a promising approach. This would enable the usage of those methods also for users who are not experts in operations research and statistics.
- **Integration in industrial data environment:** Up to now the simulation based concept is based on data gathered manually from other sources (e.g. production data acquisition systems, maintenance data) respectively with one

time measurements on technical equipment (e.g. energy consumption). Without a doubt a continuous data exchange through embedment of the simulation approach into the company`s data environment would bear potential for improvement. A coupling with e.g. applications for *energy monitoring and/or production data acquisition* could provide the simulation with the latest data and improve the consistency with the real manufacturing system. With an interface with *ERP (enterprise resource planning) systems* e.g. alternative scenarios for the planning of production capacities could be automatically evaluated with the simulation in order to provide decision support. Thinking even further, coupling with *machine control* might be beneficial as from system perspective suggestions for ideal modes of operation can be given or even automatically triggered.

- **Coupling with process simulation:** For most manufacturing processes specific process simulation applications are available which allow the detailed analysis on a physical level (e.g. FEM simulation in machining or die casting). This is not the focus of the proposed simulation approach but a coupled consideration with those applications is technically possible and might be advantageous (e.g. Herrmann et. al. 2011b). Whereas the manufacturing system simulation considers the process as black box with inputs and outputs, with a detailed physical model the actual impact of certain variables determined on system level can be realistically assessed and useful information can be generated. As an example, fluctuating process times or volatile supply with heat might influence the quality of the product.

References

Abernethy, R. B. (2006): The new Weibull handbook - Reliability & statistical analysis for predicting life, safety risk, support costs failures and forecasting warranty claims. 5. ed., North Palm Beach, Fla.: R.B. Abernethy.

Anderson, D. R. (2002): Statistics for business and economics. 8. ed., Cincinnati, Ohio: South-Western.

Andreassi, L.; Ciminelli, M. V.; Feola, M.; Ubertini, S. (2009): Innovative method for energy management: Modelling and optimal operation of energy systems. Energy and Buildings, 41/4, pp. 436–444.

Arnold, D. (2002): Handbuch Logistik - mit 77 Tabellen. VDI-Buch, Berlin: Springer.

Babcock and Wilcox Company (2010): Steam Its Generation and Use. Kessinger Publishing.

Banks, J. (2010): Discrete-event system simulation. 5. ed., internat. ed., Upper Saddle River, NJ: Pearson.

Barbian, P. (2005): Produktionsstrategie im Produktlebenszyklus - Konzept zur systematischen Umsetzung durch Produktionsprojekte. Dissertation, Techn. Univ. Kaiserslautern.

Bayer, J. (Ed.) (2003): Simulation in der Automobilproduktion. Berlin: Springer.

Bertsche, B. ; Lechner, G. (2004): Zuverlässigkeit im Fahrzeug- und Maschinenbau - Ermittlung von Bauteil- und System-Zuverlässigkeiten. 3. ed., Berlin, Heidelberg: Springer.

Beyene, A. (2005): Energy Efficiency and Industrial Classification. Energy Engineering Journal, 102/2, pp. 59–80.

Bickford, D.; Leong, K.; Ward, P. (1996): Configurations of Manufacturing Strategy, Business Strategy, Environment and Structure. Journal of Management, 22/4, pp. 597–626.

Bierbaum, U.; Hütter, J. (2004): Druckluft-Kompendium. 6. ed., Darmstadt: Hoppenstedt Bonnier Zeitschriften.

Binding, H. J. (1988): Grundlagen zur systematischen Reduzierung des Energie- und Materialeinsatzes (Bases for the systematic reduction of the use of energy and material). Dissertation, RWTH Aachen.

Birolini, A. (2010): Reliability Engineering - Theory and Practice. Berlin, Heidelberg: Springer.

Black, K. (2008): Business statistics - For contemporary decision making. 5. ed., Hoboken, NJ: Wiley.

BMWi (2011): German Federal Ministry of Economics and Technology (BMWi) - Energy Statistics, Available online: www.bmwi.de.

Bode, H.-O. (2007): Einfluss einer energieeffizienten Produktion auf Planungs- und Produktprämissen am Beispiel der Motorenfertigung. In: Fraunhofer IPK (Eds.): XII. Internationales Produktionstechnisches Kolloquium, Berlin, pp. 299–305.

Böge, A. (Ed.) (2009): Handbuch Maschinenbau - Grundlagen und Anwendungen der Maschinenbau-Technik. Wiesbaden: Vieweg+Teubner, available online http://dx.doi.org/10.1007/978-3-8348-9249-2.

Bonneschky, A. (2002): Energiekennzahlen in PPS-Systemen, Diss. u.d.T.: Bonneschky, A.: Integration energiewirtschaftlicher Aspekte in Systeme der Produktionsplanung und –steuerung. Dissertation, Technische Universität Cottbus.

Boos, W.; Kuhlmann, K. (2010): Bewertung der Ressourceneffizienz von Werkzeugmaschinen. wt Werkstattstechnik, 100/5, pp. 350–353.

Borshchev, A.; Filippov, A. (2004): From system dynamics and discrete event to practical agent based modeling: Reasons, techniques, tools. In: Proceedings of the 22nd International Conference of the System Dynamics Society, Oxford, UK.

Brans, J.; Mareschal, B.; Vincke, P. (1986): How to select and how to rank projects: The PROMETHEE method for MCDM. International Journal of Operations Research, 24, pp. 228–238.

Brettar, T. (1988): Konzeption einer betrieblichen Energiewirtschaft. Dissertation, Universität Saarbrücken, Betriebswirtschaftliche Beiträge zu Energie, Rohstoff und Umweltfragen Nr. 5, Frankfurt am Main: Lang.

Brigham, E. F.; Houston, J. F. (2009): Fundamentals of financial management. 12. ed., Mason, Ohio: South-Western Cengage Learning.

Brüggemann, A. (2005): KFW-Befragung zu den Hemmnissen und Erfolgsfaktoren von Energieeffizienz in Unternehmen. Publikation der Volkswirtschaftlichen Abteilung, Frankfurt, Main: KfW.

Brundtland Commission (1987): Our common future. Oxford paperbacks, Oxford: Oxford University Press.

C. I. R. P. (2004a): Wörterbuch der Fertigungstechnik / Dictionary of Production Engineering /Dictionnaire des Techniques de Production Mechanique Vol. III - Produktionssysteme/Manufacturing Systems/Systemes de Production. Berlin: Springer.

C. I. R. P. (2004b): Wörterbuch der Fertigungstechnik/Dictionary of Production Engineering/Dictionnaire des Techniques de Production Mechanique Vol. II - Trennende Verfahren/Material Removal Processes/Procedes d'enlevement de matiere. Berlin: Springer.

C. I. R. P. (2008): CIRP Unified Terminology of Manufacturing Systems. Berlin: Springer

Cannata, A.; Karnouskos, S.; Taisch, M. (2010): Energy efficiency driven process analysis and optimization in discrete manufacturing. Industrial Electronics, IECON'09 - 35th Annual Conference of IEEE.

Cassandras, C. G.; Lafortune, S. (2008): Introduction to Discrete Event Systems. New York: Springer, available online http://dx.doi.org/10.1007/978-0-387-68612-7.

Chadderton, D. (2004): Building Services Engineering. Taylor & Francis.

Chiotellis, C.; Seliger, G.; Weinert, N. (2009): Energy-aware Production Planning and Control. In: Proceedings of 16th CIRP International Conference on Life Cycle Engineering (LCE 2009), Cairo, Egypt, pp. 310–315.

Chow, W. K. (1996): Application of Computational Fluid Dynamics in building services engineering. Building and Environment, 31/5, pp. 425-436.

Chung, C. A. (2004): Simulation modeling handbook - A practical approach. Industrial and manufacturing engineering series, Boca Raton, Fla.: CRC Press.

Clarke, C.; Bras, B.; Guldberg, T.; Nader, G. (2008): Future Manufacturing Energy Networks. In: Proceedings of 15th CIRP International Conference on Life Cycle Engineering (LCE 2008), Sydney, Australia, pp. 420–425.

Cohen, J. (2009): Statistical power analysis for the behavioral sciences. 2. ed., New York: Psychology Press.

Dahmus, J.; Gutowski, T. (2004): An environmental analysis of machining. ASME International Mechanical Engineering Congress and RD&D Exposition, Anaheim, California, USA.

Dalzell, J. M. (2000): Food industry and the environment in the European Union - Practical issues and cost implications. 2. ed., Gaithersburg, Md.: Aspen Publ.

Dehli, M. (1998): Energieeinsparung in Industrie und Gewerbe - Praktische Möglichkeiten des rationellen Energieeinsatzes in Betrieben. Kontakt & Studium Nr. 535, Renningen-Malmsheim: expert-Verl.

DIN 8580 (2003): Fertigungsverfahren. Berlin: Beuth.

DIN EN 16001 (2009): Energiemanagementsysteme. Berlin: Beuth.

DIN 9000 (2009): Qualitätsmanagement. Berlin: Beuth.

Devoldere, T.; Dewulf, W.; Deprez, W.; Duflou, J. R. (2008): Energy Related Life Cycle Impact and Cost Reduction Opportunities in Machine Design: The Laser Cutting Case. In: Proceedings of 15th CIRP International Conference on Life Cycle Engineering (LCE 2008), Sydney, Australia, pp. 412–419.

Devoldere, T.; Dewulf, W.; Deprez, W.; Willems, B.; Duflou, J. R. (2007): Improvement Potential for Energy Consumption in Discrete Part Production Machines. In: Takata, S.; Umeda, Y. (Eds.): Advances in Life Cycle Engineering for Sustainable Manufacturing Businesses - Proceedings of the 14th CIRP Conference on Life Cycle Engineering, Waseda University, Tokyo, Japan, London: Springer-Verlag, pp. 311–316.

Dietmair, A.; Verl, A. (2008): Zustandsbasierte Energieverbrauchsprofile. wt Werkstattstechnik, 98-7/8, pp. 640–645.

Dietmair, A.; Verl, A. (2009): A generic energy consumption model for decision making and energy efficiency optimisation in manufacturing. International Journal of Sustainable Engineering, 2/2, pp. 123–133.

Dietmair, A.; Verl, A. (2010): Energy Consumption Assessment and Optimisation in the Design and Use Phase of Machine Tools. In: Proceedings of the 17th CIRP International Conference on Life Cycle Engineering (LCE 2010), Hefei, China, pp. 116–122.

Dyckhoff, H. (1994): Betriebliche Produktion - Theoretische Grundlagen einer umweltorientierten Produktionswirtschaft. Berlin: Springer.

Dyckhoff, H.; Spengler, T. S. (2010): Produktionswirtschaft - Eine Einführung. Berlin, Heidelberg: Springer.

Dyckhoff, H.; Souren, R. (2008): Nachhaltige Unternehmensführung - Grundzüge industriellen Umweltmanagements. Berlin: Springer.

Eckebrecht, J. (2000): Umweltverträgliche Gestaltung von spanenden Fertigungsprozessen, Aachen, Germany: Shaker Verlag.

Effenberger, H. (2000): Dampferzeugung, VDI-Buch, Berlin: Springer.

EIA (2009): U.S. Energy Information Administration. Available online: http://www.eia.doe.gov/.

Eichhammer, W.; Jochem, E.; Patel, M.; Tönsing, E. (1996): Overview of Energy RD&D Options for a Sustainable Future. European Commission, Luxembourg.

Einstein, D.; Worrell E.; Khrushch M. (2001): Steam systems in industry: Energy use and energy efficiency improvement potentials. In: Proceedings of the 2001 ACEEE Summer study on Energy Efficiency in Industry, pp. 535–548.

Energy Star (2010): Website. Available online: http://www.energystar.gov.

Engelhardt-Nowitzki, C.; Krenn, B.; Nowitzki, O. (Eds.) (2008): Praktische Anwendung der Simulation im Materialflussmanagement - Erfolgsfaktoren und Implementierungsszenarien. 1. Ed., Gabler Edition WissenschaftLeobener Logistik Cases, Wiesbaden: Gabler Verlag / GWV Fachverlage GmbH Wiesbaden.

Engelmann, J. (2009): Methoden und Werkzeuge zur Planung und Gestaltung energieeffizienter Fabriken.

European Comission (2006): Action Plan for Energy Efficiency: Realising the Potential. Brussels.

European Comission (2008): Energy Efficiency in Manufacturing - The Role of ICT. ICT and Energy Efficiency - Consultation Group on Smart Manufacturing. Available online: doi 10.2759/43414.

European Comission (2009): Reference Document on best available techniques for energy efficiency. Available online: http://eippcb.jrc.es/reference/.

EuP Directive 2005/32/EC (2005): Establishing a framework for the setting of ecodesign requirements for energy-using products (EuP Directive). European Parliament.

Federal Ministry for the Environment (2010): DIN EN 16001: Energy Management Systems in Practice - A Guide for Companies and Organisations, Umweltbundesamt, Berlin.

Ferretti, I.; Zanoni, S.; Zavanella, L. (2008): Energy efficiency in a steel plant using optimization-simulation. In: Proceedings of 20th European Modeling & Simulation Symposium.

Fichter, K. (2005): Interpreneurship - Nachhaltigkeitsinnovationen in interaktiven Perspektiven eines vernetzenden Unternehmertums, Habilitation, Universität Oldenburg, Theorie der Unternehmung Nr. 33, Marburg: Metropolis-Verl.

Fiedler, T.; Metz, D.; Ott, S. (2007): Künstliche Neuronale Netze zur Lastprognose im Strom- und Gasbereich. Querschnitt - Magazin der Hochschule Darmstadt, 2, pp. 135–138.

Fleschutz, T.; Rahman, A. A. A.; Harms, R.; Seliger, G. (2010): Assessment of Life Cycle Impacts and Integrated Evaluation Concept for Equipment Investment. In: Proceedings of 17th CIRP International Conference on Life Cycle Engineering (LCE 2010), Hefei, China, pp. 83–87.

Forrester, J. W. (1969): Urban Dynamics. Taylor & Francis Inc.

Fraunhofer IWU (2008): Energieeffizienz in der Produktion - Untersuchung zum Handlungs- und Forschungsbedarf. München.

Freeman, S. L.; Niefer, M. J.; Roop, J. M. (1996): Measuring Industrial Energy Eficiency: Physical Volume Versus Economic Value. Report to the U.S. Department of Energy.

Gäbel, K.; Tillman, A.-M. (2005): Simulating operational alternatives for future cement production. Journal of Cleaner Production, 13-13/14, pp. 1246–1257.

Gauchel, W. (2006): Energy-saving pneumatic systems. O + P „Ölhydraulik und Pneumatik", 50/1, pp. 1-22.

Innovationen bei der rationellen Energieanwendung (1998) Neue Chancen für die Wirtschaft, VDI-Berichte Nr. 1385, Düsseldorf: VDI-Verl.

Gloor, R. (2000): Energieeinsparungen bei Druckluftanlagen in der Schweiz. BfE-Bundesamt für Energie, Bern.

Gopalakrishnan, B. (2005): A Systems Approach to Plant-wide Energy Assessment. Energy engineering, 102/5, pp. 49-79.

Gutenberg, E. (1983): Die Produktion. 24. Ed., Enzyklopädie der Rechts- und Staatswissenschaft, Abteilung Staatswissenschaft, Berlin: Springer.

Gutowski, T. G.; Dahmus, J. B.; Thiriez, A. (2006): Electrical Energy Requirements for Manufacturing Processes. In: Duflou, J. R. (Ed.): Proceedings of the 13th CIRP Conference on Life Cycle Engineering (LCE 2006), Leuven, Belgium, pp. 623–627.

Hall, F.; Greeno, R. (2009): Building Services Handbook: Incorporating Current Building & Construction. Butterworth-Heinemann.

Heilala, J.; Vatanen, S.; Tonteri, H.; Montonen, J.; Lind, S.; Johansson, B.; Stahre, J. (2008): Simulation-based sustainable manufacturing system design. In: Mason, S. J. (Ed.): 2008 Winter Simuation Conference (WSC 2008); Miami, Florida, USA, Piscataway, NJ: IEEE, pp. 1922–1930.

Herrmann, C. (2009): Ganzheitliches Life Cycle Management: Nachhaltigkeit und Lebenszyklusorientierung in Unternehmen. Berlin: Springer.

Herrmann, C.; Thiede, S. (2009a): Process chain simulation to foster energy efficiency in manufacturing. CIRP Journal of Manufacturing Science and Technology, Vol. 1, Issue 4, pp. 221-229.

Herrmann, C.; Thiede, S. (2008): Increasing Energy Efficiency in Manufacturing Companies through Process Chain Simulation. In: Proceedings of the 6th Global Conference on Sustainable Product Development and Life Cycle Engineering, Busan, Korea, pp. 52–57.

Herrmann, C.; Thiede, S. (2009b): Towards Energy and Resource Efficient Process Chains. In: Proceedings of the 16th CIRP International Conference on Life Cycle Engineering (LCE 2009), Cairo, Egypt, pp. 303–309.

Herrmann, C.; Thiede, S.; Kara, S.; Hesselbach, J. (2011a): Energy oriented simulation of manufacturing systems - concept and application. CIRP Annals - Manufacturing Technology, Vol. 60/1, pp. 45-48.

Herrmann, C.; Heinemann, T.; Thiede, S. (2011b): Synergies from Process and Energy Oriented Process Chain Simulation – A Case Study from the Aluminium Die Casting Industry, In: Proceedings of the 18th CIRP International Conference on Life Cycle Engineering, Braunschweig, Germany, Springer, pp. 317-322.

Herrmann, C.; Thiede, S.; Heinemann, T. (2010a): Ganzheitliche Ansätze zur Erhöhung der Energie- und Ressourceneffizienz in der Produktion, In: Karlsruher Arbeitsgespräche Produktionsforschung 2010, Karlsruhe, Germany.

Herrmann, C.; Bogdanski, G.; Winter, M.; Heinemann, T.; Thiede, S.; Zein, A. (2010b): Sustainability in Production Engineering - Holistic Thinking in Education. In: Advances in Sustainable Manufacturing - Proceedings of 8th Global Conference on Sustainable Manufacturing, Abu Dhabi, VAE, pp. 25-30.

Herrmann, C.; Bergmann, L.; Thiede, S.; Halubek, P. (2007a): Total Life Cycle Management - An Integrated Approach Towards Sustainability. In: Proceedings of 3rd International Conference on Life Cycle Management, Zurich, Switzerland.

Herrmann, C.; Thiede, S.; Stehr, J.; Bergmann, L. (2008a): An environmental perspective on Lean Production. In: Proceedings of the 41th CIRP International Seminar on Manufacturing Systems, Tokyo, Japan, pp. 83–88.

Herrmann, C.; Zein, A.; Thiede, S.; Bergmann, L.; Bock, R. (2008b): Bringing sustainable manufacturing into practice: The machine tool case. In: Proceedings of the 6th Global Conference on Sustainable Product Development and Life Cycle Engineering, Busan, Korea, pp. 272–277.

Herrmann, C.; Bergmann, L.; Thiede, S.; Zein, A. (2007b): Framework for Integrated Analysis of Production Systems. In: Takata, S.; Umeda, Y. (Eds.): Advances in Life Cycle Engineering for Sustainable Manufacturing Businesses - Proceedings of the 14th CIRP Conference on Life Cycle Engineering (LCE 2007), Tokyo, Japan, London: Springer-Verlag, pp. 195–200.

Hesselbach, J.; Martin, L.; Herrmann, C.; Thiede, S.; Lüdemann, B.; Detzer, R. (2008a): Energieeffizienz durch optimierte Abstimmung zwischen Produktion und technischer Gebäudeausrüstung. In: Rabe, M. (Ed.): 13. ASIM - Fachtagung: Simulation in Produktion und Logistik /// Advances in simulation for production and logistics applications, ASIM-Mitteilung Nr. 118, Stuttgart: Fraunhofer IRB Verlag, pp. 177–185.

Hesselbach, J.; Junge, M. (2005): Reduzierung von Energiespitzen durch Fabriksimulation (Reducing Energy Peaks by Factory Simulation), Industrie Management, 21/2, pp. S. 35-37.

Hesselbach, J.; Herrmann, C.; Detzer, R.; Martin, L.; Thiede, S.; Lüdemann, B. (2008b): Energy Efficiency through optimized coordination of production and technical building services. In: Proceedings of the 15th CIRP International Conference on Life Cycle Engineering (LCE 2008), Sydney, Australia, pp. 624–629.

Hornberger, M. (o. J.): Total Energy Efficiency Management (TEEM).

Hornberger, M. (2009): Total Energy Efficiency Management - Vom Energiemanagementsystem zur Simulation der Energieverbrauchswerte auf Prozessebene. In: Elektronik ecodesign congress, München, Germany.

Hufendiek, K.; Kaltschmitt, M. (1998): Einsatz künstlicher neuronaler Netze bei der kurzfristigen Lastprognose. In: VGB PowerTech e.V. (Eds.): VGB-Konferenz, pp. 1–6.

ies – Industrie Engineering Service GmbH (2009): Corporate Website. Available online: http://strahlen-mit-trockeneis.com/ServiceProdukte.

IHK (2009): Energiepreise und Unternehmensentwicklung in Baden-Württemberg, Auswertung einer Umfrage der Industrie- und Handelskammern Heilbronn-Franken, Hochrhein-Bodensee, Karlsruhe and Ostwürttemberg, autumn 2008.

ISO EN15603 (2008): Energy performance of buildings. Overall energy use and definition of energy ratings.

Jaffe, A. B.; Stavins, R. N. (1994): The energy efficiency gap: what does it mean? Energy Policy, 22/10, pp. 60–71.

Jahangirian, M.; Eldabi, T.; Naseer, A.; Stergioulas, L. K.; Young, T. (2010): Simulation in manufacturing and business: A review. European Journal of Operational Research, 203, pp. 1–13.

Johansson, B.; Mani, M.; Skoogh, A.; Leong, S. (2009a): Discrete Event Simulation to generate Requirements Specification for Sustainable Manufacturing Systems Design. In: Proceedings of PerMIS '09 Proceedings of the 9th Workshop on Performance Metrics for Intelligent Systems.

Johansson, B.; Kacker, R.; Kessel, R.; McLean, C.; Sriram, R. (2009b): Utilizing Combinatorial Testing on Discrete Event Simulation Models for Sustainable Manufacturing. In: Proceedings of 2009 ASME Design for Manufacturing and the Life Cycle Conference, San Diego, California, USA, pp. 1095-1101.

Junge, M. (2007): Simulationsgestützte Entwicklung und Optimierung einer energieeffizienten Produktionssteuerung, Dissertation, Universität Kassel, Produktion & Energie, Vol. 1, Kassel.

Kaiser, S.; Starzer, O. (1999): Handbuch für betriebliches Energiemanagement. Energieverwertungsagentur.

Kelton, W. D.; Sadowski, R. P.; Swets, N. B. (2010): Simulation with Arena. 5. ed., Boston, Mass.: McGraw-Hill.
Khatib, H. (2003): Economic Evaluation of Projects in the Electricity Supply Industry. Institution of Electrical Engineers, Vol. 44.
Kircher, K.; Shi, X.; Patil, S.; Zhang, K. (2010): Cleanroom energy efficiency strategies: Modeling and simulation. Energy and Buildings, 42/3, pp. 282–289.
Klocke, F.; Schlosser, R.; Tönissen, S. (2010): Prozesseffizienz durch Parameterwahl - Evaluierung des Fräsprozesses. wt Werkstattstechnik, 100/5, pp. 346–349.
Krenn, B. (2007): Management komplexer Materialflüsse mittels Simulation: State-of-the-Art und innovative Konzepte: Duv.
Kuhrke, B.; Schrems, S.; Eisele, C.; Abele, E. (2010): Methodology to assess the energy consumption of cutting machine tools. In: Proceedings of the 17th CIRP International Conference on Life Cycle Engineering (LCE 2010), Hefei, China, pp. 76–82.
Lang, B.; Hesselbach, J. (2009): Energiebedarfsvorhersage produzierender Unternehmen mithilfe Neuronaler Netze. In: Gnauck, A.; Luther, B. (Eds.): ASIM 2009 - 20. Symposium Simulationstechnik, pp. 132–136.
Lanz, M. S.; Mani, M.; Leong, S. K.; Lyons, K. W.; Ranta, A.; Ikkala, K.; Bengtsson, N. (2010): Impact of Energy Measurements in Machining Operations. In: Proceedings of the ASME 2010 International Design Engineering Technical Conferences & Computers and Information in Engineering Conference, Montreal, Quebec, Canada, pp. 1-7.
Larsson, M.; Dahl, J. (2003): Reduction of the Specific Energy Use in an Integrated Steel Plant — The Effect of an Optimisation Model. ISIJ International, 10, pp. 1664–1673.
Lau, H. C. W.; Cheng, E. N. M.; Lee, C. K. M.; Ho, G. T. S. (2007): A fuzzy logic approach to forecast energy consumption change in a manufacturing system. Expert Systems with Applications, 34/3, pp. 1813-1824.
Law, A. M. (2007): Simulation modeling and analysis. 4. ed., Boston: McGraw-Hill.
Lind, S.; Johansson, B.; Stahre, J.; Berlin, C.; Fasth, Å.; Heilala, J.; Helin, K.; Kiviranta, S.; Krassi, B.; Montonen, J. (2009): SIMTER-A Joint Simulation Tool for Production Development.
Löfgren, B. (2009): Capturing the life cycle environmental performance of a company's manufacturing system, diploma thesis, Chalmers University of Technology.
Mark, K.; Rojíék, J.; Stluka, P. (2009): The Role of Predictive Models in Energy Efficiency Optimization of Complex Industrial Plants, Chemical Engineering, 18.
Massachusetts Institute of Technology (MIT) (2010): What is Operations Management? Available online: http://mitsloan.mit.edu/omg/om-definition.php.
McKinney, M. L.; Schoch, R. M.; Yonavjak, L. (2007): Environmental science - Systems and solutions. 4. ed., Sudbury, Mass: Jones and Bartlett.
Müller, E.; Engelmann, J.; Löffler, T.; Strauch, J. (2009): Energieeffiziente Fabriken planen und betreiben, Berlin, Heidelberg: Springer.
Müller, L. (2001): Handbuch der Elektrizitätswirtschaft - Technische, wirtschaftliche und rechtliche Grundlagen. 2. Ed., Elektrische Energietechnik, Berlin: Springer.
Neumann, K. (1985): Betriebliche Umweltschutzplanung mit Hilfe der Simulation - Ein integrierter Planungsansatz mit Anwendung auf einen landwirtschaftlichen Betrieb. Frankfurt am Main, New York: P. Lang.
Offner, K. (2001): Betriebliches Energiemanagement - Qualitätsmerkmale - Lieferantenfestlegung – Qualitätstechniken. Dissertation, Technische Universität Graz, 1. Ed., Techno-ökonomische Forschung und Praxis, Wiesbaden: DUV Dt. Univ.-Verl.

Östergren, K.; Janestad, H.; Berlin, J.; Sonesson, U. (2008): Integrated Simulation Technology for safe sustainable Food Processes. In: European Technology Institute - EUROSIS: FOODSIM 2008, pp. 181–185.

Oxford University Press (2011): Oxford Dictionaries Online. Available online: http://www.oxforddictionaries.com.

Patterson, M. G. (1996): What is energy efficiency? - Concepts, indicators and methodological issues. Energy Policy, 24/5, pp. 377–390.

Pegden, C. D.; etc.; Sadowski, R. P.; Shannon, R. E. (1995): Introduction to Simulation Using Siman. McGraw-Hill Education - Europe.

Pinero, E. (2009): ISO 50001: Setting the Standard for Industrial Energy Management. Green Manufacturing News.

Planck, M.; Päsler, M. (1964): Vorlesungen über Thermodynamik. 11. Ed., Berlin: de Gruyter.

Pritsker, A. A. B. (1995): Introduction to simulation and SLAM II. 4. ed., New York: Wiley.

Quadriguasi, J.; Walther, G.; Bloemhof, J.; van Nunen, J. A. E. E.; Spengler, T. (2009): A methodology for assessing eco-efficiency in logistics networks. European Journal of Operational Research, 193/3, pp. 670-682.

Rabe, M.; Spiekermann, S.; Wenzel, S. (2008): Verifikation und Validierung für die Simulation in Produktion und Logistik - Vorgehensmodelle und Techniken. VDI-Buch, Berlin, Heidelberg: Springer.

Radgen, P.; Blaustein, E. (Eds.) (2001): Compressed Air Systems in the European Union: Energy, Emissions, Savings Potential and Policy Actions. Final report, Stuttgart: LOG_X Verlag.

Rager, M. (2008): Energieorientierte Produktionsplanung - Analyse, Konzeption und Umsetzung. Dissertation, Universität Augsburg, Gabler Edition Wissenschaft, Wiesbaden: Gabler.

Rahimifard, S.; Seow, Y.; Childs, T. (2010): Minimising Embodied Product Energy to support energy efficient manufacturing. CIRP Annals - Manufacturing Technology, 59/1, pp. 25–28.

Rahimifard, S. (2009): Minimisation of Energy Consumption during the Manufacturing Phase of a Product Life Cycle. In: Manufuture 2009, Göteborg.

Rajan, G. G. (2008): Optimizing Energy Efficiencies in Industry: McGraw-Hill Education - Europe.

Rebhan, E. (Ed.) (2002): Energiehandbuch - Gewinnung, Wandlung und Nutzung von Energie. Engineering online library, Berlin: Springer.

Recknagel, H.; Sprenger, E.; Schramek, E.-R. (Eds.) (2010): Taschenbuch für Heizung + Klimatechnik 11/12 - Komplettversion: Oldenbourg Industrieverlag.

Ridder, N. (2003): Öffentliche Energieversorgungsunternehmen im Wandel - Wettbewerbsstrategien im liberalisierten deutschen Strommarkt. Marburg: Tectum-Verl.

Ruppelt, E. (Ed.) (2003): Druckluft-Handbuch. 4. Ed., Essen: Vulkan-Verl.

Saacke (2009): Faustformelsammlung. 6. Ed., Bremen.

Saidur, R.; Rahim, N. A.; Hasanuzzaman, M. (2010): A review on compressed-air energy use and energy savings. Renewable and Sustainable Energy Reviews, 14/4, pp. 1135–1153.

Sankey, H. R. (1898): Introductory note on the thermal efficiency of steam-engines. Minutes of Proceedings of The Institution of Civil Engineers., pp. 278–283.

Schenk, M. (2004): Fabrikplanung und Fabrikbetrieb - Methoden für die wandlungsfähige und vernetzte Fabrik. Berlin, Heidelberg: Springer.

Schieferdecker, B. (2006): Energiemanagement-Tools - Anwendung im Industrieunternehmen. Berlin, Heidelberg: Springer.

Schmid, C. (2008): Energieeffizienz in Unternehmen - Eine wissensbasierte Analyse von Einflussfaktoren und Instrumenten. ETH Zürich: vdf Hochschulverlag AG.

Schmid, C.; Layer, G. (2003): Möglichkeiten, Potenziale, Hemmnisse und Instrumente zur Senkung des Energieverbrauchs branchenübergreifender Techniken in den Bereichen Industrie und Kleinverbrauch. Umweltforschungsplan des Bundesministeriums für Umwelt, Naturschutz und Reaktorsicherheit, Rationelle Energieerzeugung und -nutzung, Karlsruhe: ISI.

Schmidt, J. W. (1970): Simulation and analysis of industrial systems, Homewood, Ill.: Irwin.

Schmidt, M. (2008a): The Sankey Diagram in Energy and Material Flow Management - Part I History. Journal of industrial ecology, 12/1, pp. 82-94.

Schmidt, M. (2008b): The Sankey Diagram in Energy and Material Flow Management - Part II Methodology and Current Applications. Journal of industrial ecology, 12/2, pp.173-185.

Schmidt, M. (2006): Der Einsatz von Sankey-Diagrammen im Stoffstrommanagement, Beiträge der Hochschule Pforzheim Nr. 124, Pforzheim: Hochsch.

Schmidt, M. (2007): Rohstoffe und Ressourceneffizienz - bereit für den Wettbewerb um Nachhaltigkeit?, In: "Statussemeinar 2007", Berlin. Available online: http://www.netzwerk-lebenszyklusdaten.de/cms/webdav/site/lca/shared/Veranstaltungen/Statusseminar/Statusseminar_2007/Presentations/04_Schmidt_NetLZD.pdf.

Schröter, M.; Weißfloch, U.; Buschak, D. (2009): Energieeffizienz in der Produktion – Wunsch oder Wirklichkeit? Energieeinsparpotenziale und Verbreitungsgrad energieeffizienter Techniken, Fraunhofer-Institut für System- und Innovationsforschung ISI, Karlsruhe.

Schufft, W. (2007): Taschenbuch der elektronischen Energietechnik. Hanser Fachbuchverlag.

Schuh, G. (2006): Produktionsplanung und -steuerung - Grundlagen, Gestaltung und Konzepte. VDI-Buch, Available online: http://dx.doi.org/10.1007/3-540-33855-1 / http://www.gbv.de/dms/hebis-darmstadt/toc/177220961.pdf.

Schultz, A. (2002): Methode zur integrierten ökologischen und ökonomischen Bewertung von Produktionsprozessen und -technologien, Dissertation, Universität Magdeburg.

Scottish government (2006): Scottish Energy Study, http://scotland.gov.uk.

Seefeldt, F.; Wünsch, M. (2007): Potenziale für Energieeinsparung und Energieeffizienz im Lichte aktueller Preisentwicklungen. Prognos AG, Final report 18/06.

Siemens AG (2010): Simulation of the Energy Consumption of Conveyor Lines with Tecnomatix Plant Simulation, Siemens AG. Available online: slideshare.net/ SiemensPLM, last update 2010.

Solding, P.; Petku, D.; Mardan, N. (2009): Using simulation for more sustainable production systems – methodologies and case studies. International Journal of Sustainable Engineering, 2/2, pp. 111–122.

Solding, P.; Thollander, P. (2006): Increased energy efficiency in a Swedish iron foundry through use of discrete event simulation. In: Proceedings of the 2006 Winter Simulation Conference (WSC 06), Monterey, California, USA, Piscataway, NJ: IEEE Operations Center, pp. 1971–1976.

Solding, P.; Petku, D. (2005): Applying Energy Aspects on Simulation of Energy-intensive Production Systems. In: Proceedings of the 2005 Winter Simulation Conference, Orlando, Florida, U.S.A, pp. 1428–1432.

Soper, D. S. (2011): A-priori Sample Size Calculator for Student's t-Test. Department of Information Systems and Decision Sciences, California State University. available online: http://www.danielsoper.com/statcalc/calc47.aspx.

Sorrell, S.; Schleich, J.; Scott, S.; O'Malley, E.; Trace, F.; Boede, U.; Ostertag, K.; Radgen, P. (2000): Barriers to Energy Efficiency in Public and Private Organisations. Final report to the European Commission.

Specht, H. (2005): Stromlieferverträge im liberalisierten Energiemarkt - Gestaltung von Sonderverträgen und Ausschreibung von Stromlieferungen. Berlin: VDE-Verl.

Spirax Sarco (2006): Grundlagen der Dampf- und Kondensattechnologie. Konstanz.

Spirax-Sarco Limited (2005): The Steam and Condensate Loop. Spirax-Sarco Limited.

Sterman, J. (2000): Business Dynamics: Systems Thinking and Modeling for a Complex World. McGraw-Hill Education - Europe.

Stern, P. C. (1984): Energy use - The human dimension. New York: Freeman.

Thamling, N.; Seefeldt, F.; Glöckner, U. (2010): Rolle und Bedeutung von Energieeffizienz und Energiedienstleistungen in KMU. Prognos AG, Basel.

Thiede, S.; Herrmann, C. (2010): Simulation-based Energy Flow Evaluation for Sustainable Manufacturing Systems. In: Proceedings of the 17th CIRP International Conference on Life Cycle Engineering (LCE 2010), Hefei, China, pp. 99–104.

Thollander, P. (2009): Towards Increased Energy Efficiency in Swedish Industry: Barriers, Driving Forces & Policies. Dissertation, Linköping University.

Tönsing, E. (1996): Energiekostenreduzierung durch betriebliches Energiemanagement. Initiative „Energie effizient nutzen – Schwerpunkt Strom“, pp. 1–9.

U.S. Department of Energy (2010): Building Energy Software Tools Directory. Available online: http://apps1.eere.energy.gov/buildings/tools_directory.

VDI (2003): VDI 4661:2003-09 Energiekenngrößen - Definitionen - Begriffe Methodik. Berlin: Beuth.

VDI (2006): VDI 4602 Blatt 1:2006-04 Energiemanagement-Begriffe, Definitionen. Berlin: Beuth.

VDI (2007): VDI 3633 - Simulation of systems in materials handling, logistics and production. Berlin: Beuth.

Verfaillie, H. A.; Bidwell, R. (2000): Measuring eco-efficiency - A guide to reporting company performance, Dedicated to making a difference - World Business Council for Sustainable Development.

Viegas, J. (2005): Kinetic and Potential Energy: Understanding Changes Within Physical Systems. Rosen publishing group.

Wagner, W. (2010): Wasser und Wasserdampf im Anlagenbau. 2. Ed., Würzburg: Vogel Business Media.

Wahren, S. (o. J.): Total Energy Efficiency Management - Mit Energiemanagement zur Kostensenkung in der Produktion.

Waltenberger, G. (2005): Energiemanagement in der Industrie - Die energiewirtschaftlichen Grundlagen. 1. Ed., Lohmar: Eul.

Warnecke, H. J.; Westkämper, E.; Gottwald, B. (1998): Einführung in die Fertigungstechnik. Wiesbaden: Teubner.

Weinert, N. (2010): Planung energieeffizienter Produktionssysteme. ZWF, 105/5, pp. 503–507.

Weinert, N.; Chiotellis, S.; Seliger, G. (2009): Concept for Energy-Aware Production Planning based on Energy Blocks. In: Proceedings of the 7th Global Conference on Sustainable Manufacturing, pp. 75–80.

Wenzel, S.; Collisi-Böhmer, S.; Pitsch, H.; Rose, O.; Weiß, M. (2008): Qualitätskriterien für die Simulation in Produktion und Logistik - Planung und Durchführung von Simulationsstudien. VDI-Buch, Berlin, Heidelberg: Springer.

Westerkamp, T. A. (2008): It's not easy being green - energy management appraises the details. Industrial Engineer, 3, pp. 37–41.

Westkämper, E. (2005): Einführung in die Organisation der Produktion. Berlin: Springer.

Wischhusen, S.; Lüdemann B.; Schmitz G. (2003): Economical Analysis of Complex Heating and Cooling Systems with the Simulation Tool HKSim. In: Proceedings of the 3rd International Modelica Conference, Linköping, Sweden.

Wohinz, J. W.; Moor, M. (1989): Betriebliches Energiemanagement - Aktuelle Investition in die Zukunft. Wien: Springer.

Wohlgemuth, V.; Page, B.; Kreutzer, W. (2006): Combining discrete event simulation and material flow analysis in a component-based approach to industrial environmental protection. Environmental Modelling & Software, 21/11, pp. 1607–1617.

Wohlgemuth, V. (2005): Komponentenbasierte Unterstützung von Methoden der Modellbildung und Simulation im Einsatzkontext des betrieblichen Umweltschutzes - Konzeption und prototypische Entwicklung eines Stoffstromsimulators zur Integration einer stoffstromorientierten Perspektive in die auftragsbezogene Simulationssicht. Dissertation, Universität Hamburg, Aachen: Shaker.

Yager, R. R.; Zadeh, L. A. (1992): An Introduction to Fuzzy Logic Applications in Intelligent Systems: Kluwer Academic Publishers.

Zeigler, B. P.; Praehofer, H.; Kim, T. G. (2000): Theory of modeling and simulation. Academic press New York, NY.

Own References

2011

Herrmann, C.; Thiede, S.; Kara, S.; Hesselbach, J. (2011): Energy oriented simulation of manufacturing systems - concept and application, In: *CIRP Annals - Manufacturing Technology*: Elsevier, Vol. 60, Issue 1, pp. 45-48.

Herrmann, C.; Kara, S.; Thiede, S. (2011): Dynamic life cycle costing based on lifetime prediction. *International Journal of Sustainable Engineering*, Vol. 4, Issue 3, pp. 224-235.

Thiede, S.; Herrmann, C.; Kara, S. (2011): State of research and an innovative approach for simulating energy flows of manufacturing systems, In: Proceedings of the *18th CIRP International Conference on Life Cycle Engineering*, Braunschweig, Germany, Springer, pp. 335-340.

Herrmann, C.; Heinemann, T.; Thiede, S. (2011): Synergies from Process and Energy Oriented Process Chain Simulation – A Case Study from the Aluminium Die Casting Industry, In: Proceedings of the *18th CIRP International Conference on Life Cycle Engineering*, Braunschweig, Germany, Springer, pp. 317-322.

2010

Thiede, S.; Herrmann, C. (2010): Simulation-based Energy Flow Evaluation for Sustainable Manufacturing Systems, In: Proceedings of the *17th CIRP International Conference on Life Cycle Engineering*, Hefei, China, pp. 99-104.

Herrmann, C.; Bogdanski, G.; Winter, M.; Heinemann, T.; Thiede, S.; Zein, A. (2010): Sustainability in Production Engineering - Holistic Thinking in Education, In: Advances in Sustainable Manufacturing - Proceedings of *8th Global Conference on Sustainable Manufacturing*, Abu Dhabi, pp. 25-30.

Thiede, S., Herrmann, C. (2010): Energy Flow Simulation for Manufacturing Systems, In: Advances in Sustainable Manufacturing - Proceedings of *8th Global Conference on Sustainable Manufacturing*, Abu Dhabi, pp. 275-280.

Herrmann, C.; Thiede, S.; Heinemann, T. (2010): A Holistic Framework for Increasing Energy and Resource Efficiency in Manufacturing, In: Advances in Sustainable Manufacturing - Proceedings of *8th Global Conference on Sustainable Manufacturing*, Abu Dhabi, pp. 267-273.

Bergmann, L.; Thiede, S.; Herrmann, C. (2010): Mehr Energie- und Ressourceneffizienz in der Produktion, In: *Nachhaltige Produktion*, SellersMedia, 2010, Issue 2, pp. 30-32, ISSN 1868-4181.

Herrmann, C.; Heinemann, T.; Thiede, S. (2010): Identifying levers for Enhancing Energy and Resource Efficiency in Industrial Process Chains using the example of Aluminium Die Casting: Proceedings of the *1st International Conference on Automotive Materials and Manufacturing*, Pune, India, pp. 119–129.

Herrmann, C.; Kara, S.; Thiede, S.; Luger, T. (2010): Energy Efficiency in Manufacturing – Perspectives from Australia and Europe: Proceedings of the *17th CIRP International Conference on Life Cycle Engineering*, Hefei, China, pp. 23–28.

Herrmann, C.; Thiede, S.; Heinemann, T. (2010): Ganzheitliche Ansätze zur Erhöhung der Energie - und Ressourceneffizienz in der Produktion, In: Proceedings of *Karlsruher Arbeitsgespräche Produktionsforschung 2010*, Karlsruhe.

Herrmann, C.; Thiede, S.; Zein, A.; Heinemann, T. (2010): Holistic Approaches for Increasing Energy and Resource Efficiency in Manufacturing: Proceedings of the *1st International Conference on Automotive Materials and Manufacturing*, Pune, India, pp. 39–48.

2009

Herrmann, C.; Bergmann, L.; Thiede, S. (2009): Methodology for the design of sustainable production systems, *International Journal of Sustainable Manufacturing*, Vol. 1, Issue 4, pp. 376–395.

Herrmann, C.; Thiede, S. (2009): Process chain simulation to foster energy efficiency in manufacturing, *CIRP Journal of Manufacturing Science and Technology*, Vol. 1, Issue 4, pp. 221-229.

Herrmann, C.; Thiede, S. (2009): Towards Energy and Resource Efficient Process Chains, In: Proceedings of the *16th CIRP International Conference on Life Cycle Engineering*, Cairo, Egypt, pp. 303–309.

Herrmann, C.; Thiede, S.; Kuntzky, K.; Böhm, S.; Frauenhofer, M.; Gadhia, D. (2009): Indian Solar Thermal Technology: Potentials and Challenges, In: Proceedings of the *7th Global Conference on Sustainable Manufacturing*, Chennai, India, pp. 169–174.

Herrmann, C.; Thiede, S.; Luger, T.; Zein, A.; Stehr, J.; Halubek, P.; Torney, M. (2009): Automotive Life Cycle Engineering, In: Proceedings of the *16th CIRP International Conference on Life Cycle Engineering*, Cairo, Egypt, pp. 157–164.

Herrmann, C.; Thiede, S.; Zein, A.; Ihlenfeldt, S.; Blau, P. (2009): Energy Efficiency of Machine Tools: Extending the Perspective, In: Proceedings of the *42nd CIRP Conference on Manufacturing Systems*, Grenoble, Switzerland.

2008

Herrmann, C.; Bergmann, L.; Halubek, P.; Stehr, J.; Thiede, S. (2008): Life Cycle Engineering - State of the Art and Research Perspectives, In: Proceedings of the *15th CIRP International Conference on Life Cycle Engineering*, Sydney, Australia, pp. 449–454.

Herrmann, C.; Bergmann, L.; Halubek, P.; Thiede, S. (2008): Lean Production System Design from the Perspective of the Viable System Model, In: Proceedings of the *41th CIRP International Seminar on Manufacturing Systems*, Tokyo, Japan, pp. 309–314.

Herrmann, C.; Thiede, S. (2008): Increasing Energy Efficiency in Manufacturing Companies through Process Chain Simulation, In: Proceedings of the *6th Global Conference on Sustainable Product Development and Life Cycle Engineering*, Busan, Korea, pp. 52–57.

Herrmann, C.; Thiede, S.; Stehr, J.; Bergmann, L. (2008): An environmental perspective on Lean Production, In: Proceedings of the *41th CIRP International Seminar on Manufacturing Systems*, Tokyo, Japan, pp. 83–88.

Herrmann, C.; Zein, A.; Thiede, S.; Bergmann, L.; Bock, R. (2008): Bringing sustainable manufacturing into practice: the machine tool case, In: Proceedings of the *6th Global Conference on Sustainable Product Development and Life Cycle Engineering*, Busan, Korea, pp. 272–277.

Hesselbach, J.; Martin, L.; Herrmann, C.; Thiede, S.; Lüdemann, B.; Detzer, R. (2008): Energieeffizienz durch optimierte Abstimmung zwischen Produktion und technischer Gebäudeausrüstung, In: Proceedings of *13. ASIM - Fachtagung: Simulation in Produktion und Logistik - Advances in simulation for production and logistics applications*, Stuttgart: Fraunhofer IRB, pp. 177–185.

Hesselbach, J.; Herrmann, C.; Detzer, R.; Martin, L.; Thiede, S.; Lüdemann, B. (2008): Energy Efficiency through optimized coordination of production and technical building services, In: Proceedings of the *15th CIRP International Conference on Life Cycle Engineering*, Sydney, Australia, pp. 624–629.

2007

Herrmann, C.; Bergmann, L.; Thiede, S. (2007): An Integrated Approach for the Evaluation of Maintenance Strategies to Foster Sustainability in Manufacturing, In: Proceedings of *Sustainable Manufacturing V: Global Symposium on Sustainable Product Development and Life Cycle Engineering*", Rochester, NY USA.

Herrmann, C.; Bergmann, L.; Thiede, S. (2007): Developing Life Cycle Oriented Innovations Within the Turbulent Business Environment, In: Proceedings of *ICED 07 - 16th International Conference of Engineering Design*, Paris, France.

Herrmann, C.; Bergmann, L.; Thiede, S. (2007): Gestaltungselemente und Erfolgsfaktoren - Ergebnisse einer empirischen Umfrage unter produzierenden Unternehmen, *Intelligenter Produzieren*, Issue 3, pp. 20-22.

Herrmann, C.; Bergmann, L.; Thiede, S. (2007): Life Cycle Oriented Design Of Lean Production Systems, In: Proceedings of the *3rd International Virtual Design and Automation Conference*, Poznan, Poland, pp. 295–302.

Herrmann, C.; Bergmann, L.; Thiede, S.; Halubek, P. (2007): Total Life Cycle Management - An Integrated Approach Towards Sustainability, In: Proceedings of *3rd International Conference on Life Cycle Management*, Zurich, Switzerland.

Herrmann, C.; Bergmann, L.; Thiede, S.; Luger, T. (2007): Total Life Cycle Management - Framework and Concepts, In: Proceedings of *International Workshop on Sustainability in Manufacturing* - Remanufacturing for a closed-loop economy, Busan, Korea, pp. 165–176.

Herrmann, C.; Bergmann, L.; Thiede, S.; Torney, M.; Zein, A. (2007): Framework For The Dynamic And Life Cycle Oriented Evaluation Of Maintenance Strategies, In: Proceedings of the *3rd International Virtual Design and Automation Conference*, Poznan, Poland, pp. 277–284.

Herrmann, C.; Bergmann, L.; Thiede, S.; Zein, A. (2007): Energy Labels for Production Machines - An Approach to Facilitate Energy Efficiency in Production Systems, In: Proceedings of the *40th CIRP International Seminar on Manufacturing Systems*, Liverpool, UK.

Herrmann, C.; Bergmann, L.; Thiede, S.; Zein, A. (2007): Framework for Integrated Analysis of Production Systems, In: Proceedings of the *14th CIRP Conference on Life Cycle Engineering*, Tokyo, Japan: Springer, pp. 195–200.

Herrmann, C.; Bergmann, L.; Thiede, S.; Zein, A. (2007): Life Cycle Innovations in Extended Supply Chain Networks, In: Proceedings of the *14th CIRP Conference on Life Cycle Engineering*, Tokyo, Japan, Springer, pp. 439–444.

Herrmann, C.; Bergmann, L.; Thiede, S.; Zein, A. (2007): Total Life Cycle Management - A Systems and Cybernetics Approach to Corporate Sustainability in Manufacturing, In: Proceedings of *Sustainable Manufacturing V: Global Symposium on Sustainable Product Development and Life Cycle Engineering*", Rochester, NY USA.

Appendix

Steam table (taken from Spirax-Sarco Limited)

http://www.buildingdesign.co.uk/mech-technical/spirax-sarco-t2/Saturated-Steam-Tables.jpg

				Specific Enthalpy			
	Pressure		Temperature	Water	Evaporation	Steam	Specific Volume Steam
bar		kPa	C	(hf) kJ/kg	(hfg) kJ/kg	(hg) kJ/kg	m3/kg
0.3		30	69.10	289.23	2,336.10	2,625.30	5.2290
0.5	absolute	50	81.33	340.49	2,305.40	2,645.90	3.2400
0.75		75	91.78	384.39	2,278.60	2,663.00	2.2170
0.95		95	98.20	411.43	2,261.80	2,673.20	1.7770
0	gauge	0	100.00	419.04	2,257.00	2,676.00	1.6730
0.1		10	102.66	430.20	2,250.20	2,680.40	1.5330
0.2		20	105.10	440.80	2,243.40	2,684.20	1.4140
0.3		30	107.39	450.40	2,237.20	2,687.60	1.3120
0.4		40	109.55	459.70	2,231.30	2,691.00	1.2250
0.5		50	111.61	468.30	2,225.60	2,693.90	1.1490
0.7		70	115.40	484.10	2,215.40	2,699.50	1.0240
0.9		90	118.80	498.90	2,205.60	2,704.50	0.9230
1.1		110	121.96	512.20	2,197.00	2,709.20	0.8410
1.3		130	124.90	524.60	2,188.70	2,713.30	0.7730
1.5		150	127.62	536.10	2,181.00	2,717.10	0.7140
1.7		170	130.13	547.10	2,173.70	2,720.80	0.6650
1.9		190	132.54	557.30	2,166.70	2,724.00	0.6220
2.2		220	135.88	571.70	2,156.90	2,728.60	0.5680
2.6		260	140.00	589.20	2,144.70	2,733.90	0.5090
3		300	143.75	605.30	2,133.40	2,738.70	0.4610
3.4		340	147.20	620.00	2,122.90	2,742.90	0.4220
3.8		380	150.44	634.00	2,112.90	2,746.90	0.3890
4.5		450	155.55	656.30	2,096.70	2,753.00	0.3420
5.5		550	162.08	684.60	2,075.70	2,760.30	0.2920
6.5		650	167.83	709.70	2,056.80	2,766.50	0.2550
7.5		750	173.02	732.50	2,039.20	2,771.70	0.2270
8.5		850	177.75	753.30	2,022.90	2,776.20	0.2040
11		1100	188.02	798.80	1,986.00	2,784.80	0.1630
13.5		1350	196.62	837.90	1,953.20	2,791.10	0.1360
16		1600	204.38	872.30	1,923.40	2,795.70	0.1170
18.5		1850	211.25	903.10	1,895.80	2,799.00	0.1020
21		2100	217.35	931.30	1,870.10	2,801.40	0.0906
24		2400	224.02	962.20	1,840.90	2,803.10	0.0797
27		2700	230.14	990.70	1,813.30	2,804.00	0.0714
30		3000	235.78	1,017.00	1,787.00	2,804.10	0.0645
35		3500	244.26	1,057.70	1,745.50	2,803.20	0.0554
40		4000	251.94	1,094.60	1,706.30	2,800.90	0.0485
46		4600	260.13	1,135.30	1,661.60	2,796.90	0.0421
50		5000	265.26	1,160.80	1,632.80	2,793.60	0.0386

Graphical User Interface of Simulation Approach

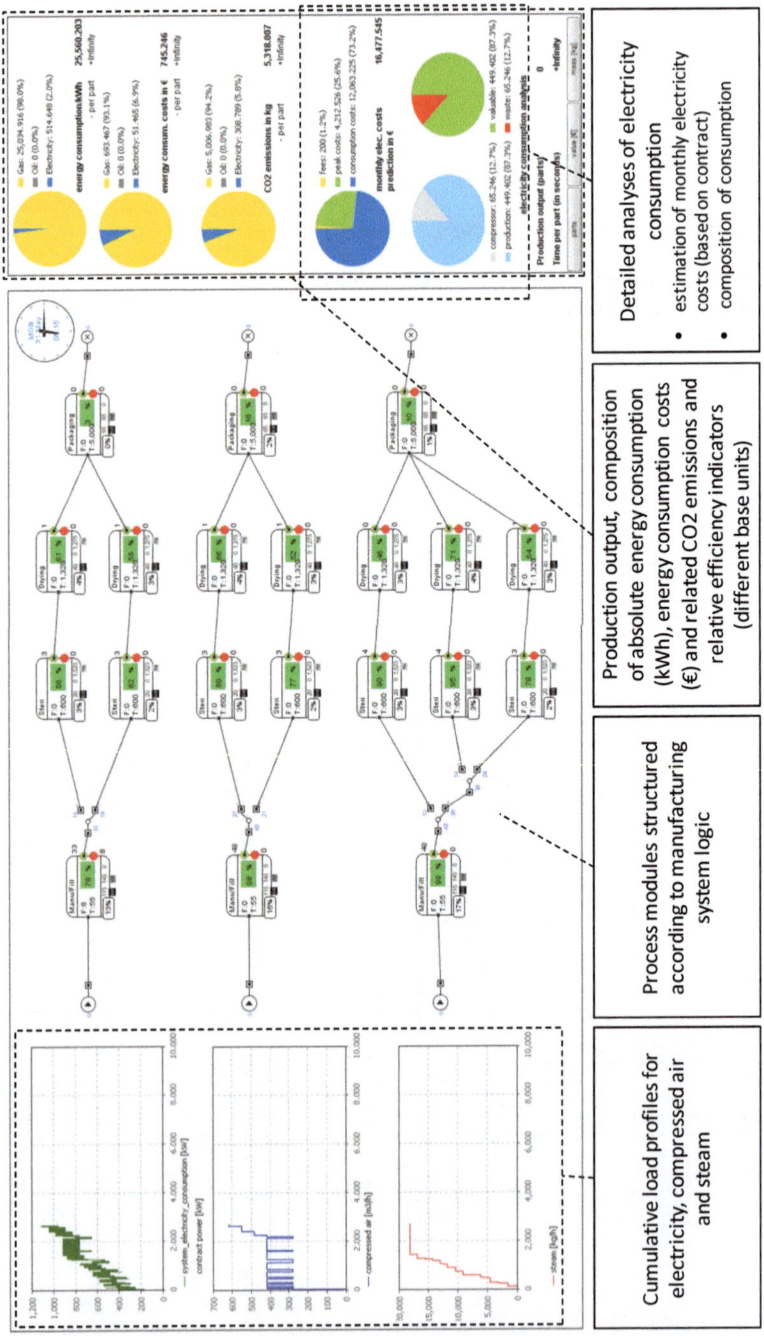

Index

MIX
Papier aus verantwortungsvollen Quellen
Paper from responsible sources
FSC® C105338

If you have any concerns about our products,
you can contact us on
ProductSafety@springernature.com

In case Publisher is established outside the EU,
the EU authorized representative is:
Springer Nature Customer Service Center GmbH
Europaplatz 3, 69115 Heidelberg, Germany

Printed by Libri Plureos GmbH
in Hamburg, Germany